T0094095

Springer Theses

Recognizing Outstanding Ph.D. Research

For further volumes:
http://www.springer.com/series/8790

Aims and Scope

The series "Springer Theses" brings together a selection of the very best Ph.D. theses from around the world and across the physical sciences. Nominated and endorsed by two recognized specialists, each published volume has been selected for its scientific excellence and the high impact of its contents for the pertinent field of research. For greater accessibility to non-specialists, the published versions include an extended introduction, as well as a foreword by the student's supervisor explaining the special relevance of the work for the field. As a whole, the series will provide a valuable resource both for newcomers to the research fields described, and for other scientists seeking detailed background information on special questions. Finally, it provides an accredited documentation of the valuable contributions made by today's younger generation of scientists.

Theses are accepted into the series by invited nomination only and must fulfill all of the following criteria

- They must be written in good English.
- The topic should fall within the confines of Chemistry, Physics, Earth Sciences and related interdisciplinary fields such as Materials, Nanoscience, Chemical Engineering, Complex Systems and Biophysics.
- The work reported in the thesis must represent a significant scientific advance.
- If the thesis includes previously published material, permission to reproduce this must be gained from the respective copyright holder.
- They must have been examined and passed during the 12 months prior to nomination.
- Each thesis should include a foreword by the supervisor outlining the significance of its content.
- The theses should have a clearly defined structure including an introduction accessible to scientists not expert in that particular field.

Ioachim Pupeza

Power Scaling of Enhancement Cavities for Nonlinear Optics

Doctoral Thesis accepted by
The University of Munich/Max Planck
Institute for Quantum Optics, Germany

 Springer

Author
Dr. Ioachim Pupeza
Max Planck Institute for Quantum
 Optics
Hans-Kopfermann-Str. 1
85748 Garching
Germany

Supervisor
Prof. Dr. Ferenc Krausz
Max Planck Institute for Quantum
 Optics
Hans-Kopfermann-Str. 1
85748 Garching
Germany

ISSN 2190-5053   ISSN 2190-5061   (electronic)
ISBN 978-1-4614-4099-4   ISBN 978-1-4614-4100-7   (eBook)
DOI 10.1007/978-1-4614-4100-7
Springer New York Heidelberg Dordrecht London

Library of Congress Control Number: 2012938352

Printed on acid-free paper

Springer is part of Springer Science+Business Media (www.springer.com)

Supervisor's Foreword

Enhancement cavities are passive optical resonators in which the power of a high repetition rate laser is increased by up to several orders of magnitude. These systems allow scaling the energy of ultrashort laser pulses and their intensity in a focus to values not accessible by oscillator-amplifier arrangements at multi-MHz repetition rates. A number of applications in science and engineering have already been realized, such as the generation of brilliant XUV-radiation, XUV frequency combs, and the emission of hard X-rays by relativistic Thomson scattering. However, to exploit the full capacity of this technique further progress is needed.

To readers not familiar with this subject it may seem surprising that energy and power of a train of short laser pulses can be increased by a factor >1,000 without interaction with an active medium. However, everything is in proper order, energy is conserved, and all physics laws are obeyed. The miraculous power boost is just the result of the sharp resonance in a high-Q optical cavity.

It turns out, however, that the realization of this concept is far from being a trivial task. For optimum overlap with the cavity mode the seed laser has to be operated in a very clean mode. To maintain the resonance, the optical length of the cavity and that of the seed laser oscillator have to be adjusted to within a fraction of a wavelength. Active stabilization is required for this purpose. Ultrahigh reflectivity mirrors with reflection coefficients exceeding 99.9 % are essential elements for a high finesse cavity. Their fabrication stretches available technological know-how to its limits. While supervising this thesis it was fascinating to see all of these pieces being put together in the laboratory.

The task assigned to this thesis work was finding and overcoming fundamental limitations in scaling the circulating power of enhancement cavities. Ioachim Pupeza succeeded outstandingly in meeting this challenge. As a main detrimental mechanism nonlinear interaction of the intense laser radiation with the cavity mirrors was identified. Third harmonic generation at the mirrors was observed as an indication of this effect. Thermal expansion of the uppermost layers of the cavity mirrors was found to be another cause contributing to degradation. Going to the limits permitted by these effects, Mr. Pupeza succeeded in generating intra-cavity laser powers which surpass previous values by one order of magnitude.

Innovative suggestions to improve the parameters even further are offered. These include novel cavity designs and the realization of shorter circulating pulses. A proposal for generating isolated attosecond-pulses using femtosecond-enhancement cavities is also included. A very preliminary experiment toward this goal is presented.

As an experimental application of the results of this work, harmonics of the fundamental 1,040 nm radiation up to the 21st order have been generated, corresponding to a wavelength of 50 nm. This radiation is coupled out by means of a Brewster window between the focus und the next mirror. A number of suggestions for further and possibly more efficient output coupling methods are presented.

Besides reporting these original results of research, a broad introduction into the physics of enhancement cavities is provided in this thesis. The basic theory, methods of measuring cavity parameters, the generation and utilization of higher transverse modes, and a review of cavity applications are included. More than 140 references enable the interested reader to obtain even deeper insight into this fascinating field of research.

Garching, Germany, 8 December 2011   Prof. Dr. Ferenc Krausz
Max-Planck-Institut für Quantenoptik
Garching
Ludwig-Maximilians-Universität München

Acknowledgments

First and foremost I would like to thank Prof. Ferenc Krausz for his continuous and tireless support which reached out far beyond technical matters and enormously boosted my learning process and my creativity in the past 4 years.

Jens Rauschenberger introduced me—as a mathematician and engineer—to the labs of experimental physics. Thanks for your patience! Furthermore, he played a leading role in initiating the projects which ensured the financial frame of my doctorate work.

I thank Ernst Fill for the great advice, founded on solid and far-reaching knowledge and for his highly motivating character. Working with you has been a great pleasure and inspiration. Also, thanks for proof-reading my thesis!

I am grateful to my colleagues Jan Kaster and Simon Holzberger, who have not only done research at my side with great enthusiasm, but have also taught me numerous things.

I owe a special "thank you!" to the fiber laser group from the IAP in Jena, who provided the laser system *Jenny* which enabled this work: Prof. Andreas Tünnermann, Prof. Jens Limpert, Tino Eidam, and Fabian Röser. Special thanks go out to Tino for the numerous visits in München, in the frame of which he repaired and upgraded Jenny. We performed many experiments and learned a lot together. Our many interesting discussions contributed a lot to my motivation in the last years. Action Jackson!

Great support came from the "Hanschies" as well. In particular, I would like to thank Prof. Hänsch, Thomas Udem, Ronald Holzwarth, Birgittta Bernhardt, Akira Ozawa, and Andreas Vernaleken. A very special "thank you!" goes out to Gitti for her unique support. She was always at my side with good advice and great help. Also a very special "thank you!" goes out to Thomas: he always found time and enthusiasm for my questions and ideas and gave me great answers and suggestions. Thank you both from teaching me so much! I also wish to thank Christoph Gohle, one of the most impressive scientists I know, for the discussions we had about physics.

One of the greatest collaborations during these years was a brief, but very pleasant and fruitful one with Xun Gu. Xun, thanks a lot for the inspiring way of tackling the problems and solving them!

I wish to thank Johannes Weitenberg and Peter Rußüldt from the Fraunhofer ILT for the many interesting discussions. In particular, Johannes keeps inspiring me again and again with his approach to science and with his passion for it. The weeks during which we carried out experiments together count among the most beautiful during my Ph.D. period and I hope that there will be many more such opportunities. I would also like to thank Dominik Esser for his fruitful efforts of drilling microscopic holes in laser mirrors.

A warm "thank you!" goes out to the colleagues from our large "Krausz-group" for their interest and support, especially Oleg Pronin, Vladimir Pervak, Alexander Apolonski, and Matthias Kling. For their active support, I would also like to thank Siegfried Herbst and Rolf Öhm from the LMU.

I would like to send special thanks to Prof. Abdallah M. Azzeer and Prof. Zeyad A. Alahmed from King Saud University in Riyadh. May our future collaborations be fruitful!

During my Ph.D. period at the MPQ and at the LMU I made many acquaintances and friendships outside the lab, while doing sports, music, philosophy, or just like that. For the beautiful time I wish to thank in particular Roswitha Graf, Catherine Teisset, Thomas Ganz, Friedrich Kirchner, Raphael Weingartner, Justin Gagnon (special thanks for the interesting discussions on quantum mechanics!), Alex Buck, Peter Jazz, Sergey Rykovanov, John Frusciante, Sarah Stebbings, Fabian Lücking, Peter Reckenthäler, Michael Hofstetter, Casey Chow, Janis Alnis. To everyone I forgot to mention in this list: sorry!

Outside the MPQ/LMU world, a special "thank you!" is due to my professors and mentors, who encouraged me to take the scientific route of the last 4 years and prepared me for it. I thank Prof. Rainer Löwen for his unique support during my studies in Mathematics and for granting me access to his way of perceiving science, in particular in the context of my diploma work in the field of incidence geometry. I thank Prof. Martin Koch for the beautiful years of THz activity in Braunschweig and for his advice to apply for this Ph.D. position. I thank Prof. Alek Kavcic for the opportunity of writing my diploma thesis in coding theory at Harvard University and for the related support. I am deeply grateful to my mentor Prof. Werner Deutsch for his wise advice and for his friendship.

Perhaps the greatest "thank you!" is due to my family who continuously gave me backing, safety, love, courage, and cheerfulness. In particular I wish to thank Susi for the love, support, and patience during the last years, my sister Vivi for the inspiration, force, and vitality, my parents: *Vom Vater hab ich die Statur/Des Lebens ernstes Führen/Vom Mütterchen die Frohnatur/Und Lust zu fabulieren*, as Goethe beautifully said. In addition, I thank my father for the continuous motivation regarding this work and for the many discussions which often lead to unique ideas. Moreover, I thank the families Rădulescu und Lera i Roura who strived to provide me with the best conditions for this doctorate. *Mulţumesc din inimă! Moltes gracies!*

Contents

Chapter 1
Introduction

Jedes Wekzeug trägt den Geist in sich, aus dem heraus es geschaffen worden ist.

Werner Heisenberg

1.1 About Light: From the Genesis to the Laser

Light is present in virtually all areas of life. It is involved in the majority of processes in nature, it plays a primordial role in most religions and it has been one of the fundamental (and perhaps the oldest) subjects of philosophy, arts and scientific research. The preoccupation with light and in particular, with the study of its origin and properties and with its control has always constituted an integral part of culture.[1]

Many principles of optics, i.e. the science of light, were already known in antiquity. Euclid's description of the principles of geometrical optics from the third century BC counts among the first documented scientific works in optics. The Greek geometrician possessed the mathematical apparatus necessary for the description of fundamental facts related to the rectilinear propagation of light. Archimedes, who lived around the same time as Euclid, is said to have defended Syracuse from the invading Roman fleet by using large mirrors to focus the sunlight on the enemy ships, setting them on fire. Another testimony of the knowledge in the field of optics during antiquity is found in a book of Ptolemy from the second century AD. In addition to his predecessor's work, he describes the phenomenon of refraction of light at interfaces between media of different optical densities.

The most important treaty on optics during the Middle Ages was written by Ibn al-Haitham (or Alhazen) around the year 1000. He delivers descriptions of the magnification of lenses, spherical aberration, parabolic mirrors and explanations for phenomena such as the formation of rainbows and atmospherical refraction.

---

[1] The following historical review is largely based on Walther's book *Was ist Licht?* [1].

I. Pupeza, *Power Scaling of Enhancement Cavities for Nonlinear Optics*,
Springer Theses, DOI: 10.1007/978-1-4614-4100-7_1,
© Springer Science+Business Media New York 2012

Together with the writings of Ptolemy, this work represents the fundamental reference on optics up to the seventeenth century. After this, the field of optics experienced rapid development, supported by the refinement of the manufacturing of lenses and measurement devices and techniques as well as by the evolution of mathematical tools. Outstanding landmarks of this development include the observation of the interference of light by R. Boyle and R. Hooke and the foundation of the wave theory for light by Ch. Huygens in the seventeenth century, experiments on interference and diffraction by Th. Young, the discovery of the polarization of light by L. Malus, detailed explanations for the interference and diffraction phenomena by Fresnel and Fraunhofer in the early nineteenth century and the direct measurement of the speed of light by L. Foucault and H. Fizeau in 1850. In the late nineteenth century, J. C. Maxwell succeeded in mathematically describing the experiments on the generation of variable electric and magnetic fields initiated 40 years earlier by M. Faraday. The speed of the electromagnetic wave propagation predicted by Maxwell's equations coincided with the previously measured speed of light. Experiments performed by H. Hertz in 1888 demonstrated electromagnetic waves directly, rounding up the classical wave picture and providing an apparent conclusion of the research on the nature and the properties of light.

However, the understanding of light-matter interaction provided by the classical wave model of light fails at describing atomic absorption and emission processes. The necessary expansion of the existing theory was provided by Max Planck in 1900 with the introduction of the quantum hypothesis. Planck postulated that an electronically oscillating system can only exchange energy of discrete amounts with an electromagnetic field. The energy E of a *quantum* of radiation is proportional to the frequency f of the radiation via $E = hf$, where *Planck's constant h* is a new fundamental physical constant. Einstein's explanation of the photoelectric effect in 1905 provided the validation of the new model, using particles of light,[2] later called *photons*. Shortly thereafter, the hypothesis of a coexistence of waves and particles was extended to electrons by L. de Broglie and supported by the observation of interference phenomena of electrons at crystal lattices by C. Davisson and L. Germer.

The wave-particle duality marks modern physics and, in particular, the field of *quantum electronics*, which studies quantum phenomena related to the interaction of electromagnetic radiation with matter. The process underlying the probably most important sources of radiation in this field is *amplification by stimulated emission of radiation*, which was first predicted and experimentally demonstrated for microwaves (therefore the acronym *maser*) in the first half of the last century, and then extended to the infrared, visible, ultraviolet and even X-ray regions of the electromagnetic spectrum, loosely designated as light, conferring the acronym *laser*.[3] The output signal of this coherent amplification process reproduces the frequency of the input

---

[2] In the second half of the seventeenth century, Newton developed a particle-based model for light. However, his theory lost ground to Huygens' wave theory, which could explain most of the experiments of that time.

[3] During the time of this doctorate work, the scientific community celebrated the 50 years anniversary of the first experimental demonstration of the laser by Th. H. Maiman in 1960.

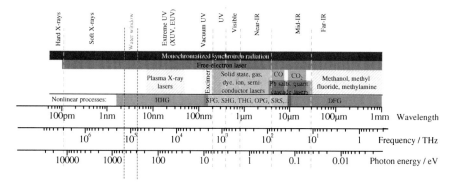

Fig. 1.1 Coherent light sources. Hashed regions: only scarce, discrete laser lines (transitions) are available. Main nonlinear processes used to convert laser light: *SHG, THG, HHG* second-, third- and high-harmonic generation, *SFG, DFG* sum- and difference-frequency generation, *OPG* optical parametric generation, *SRS* stimulated Raman scattering

electromagnetic wave with the respective phase, but with a substantially increased amplitude. In the case of the laser, the gain medium consists of a quantum-mechanical system such as a collection of atoms, molecules or ions or a semiconducting crystal, which is excited by a pumping process (providing the energy necessary for the amplification process) into higher energy levels, so that a *population inversion* is achieved.[4] In conjunction with the feedback provided by an (optical) resonator, the radiation obtained with this amplification mechanism forms a bright beam with unequaled spatial and temporal coherence properties.

The past few decades have witnessed a rapid development of laser sources, motivated by and enabling myriads of applications ranging from the investigation of microcosm dynamics over telecommunications to industrial manufacturing. Today, a great variety of gain media is available, providing quantum energy transitions for laser action over large parts of the electromagnetic spectrum between X-rays and far-infrared. Still, significant spectral regions remain, where direct laser action is either inefficient or not available at all, other than by means of large facilities such as free-electron lasers. Figure 1.1 provides an overview of the available coherent light sources, compiled using data from [3–6]. Most notably, the spectral regions below 200 nm (Vacuum UV, XUV, soft X-rays) and above 10 µm (mid- and far-IR) exhibit a sparse presence of efficient direct laser action. A convenient alternative of generating high-power coherent radiation in these spectral regions relies on the conversion of light emitted by a laser in a different spectral region via nonlinear processes. However, these processes can exhibit low conversion efficiencies and in general, their efficiency increases with increasing intensity. In the sixties, shortly after the first observation of an optical nonlinear process—namely second-harmonic generation (SHG)—the technique of passive laser light enhancement in an optical resonator was

---

[4] A detailed description of the mode of operation of the laser would exceed the frame of this section. A textbook introduction is given in Siegman's book *Lasers* [2].

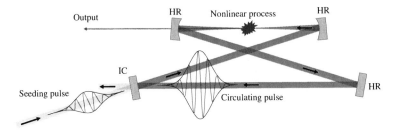

Fig. 1.2 Schematic of an enhancement cavity. *IC* input coupler, *HR* highly reflective mirror (for the fundamental radiation)

successfully employed to improve its overall conversion efficiency. In recent years, this technique has gained new importance due to the advent of high-power laser sources emitting phase-stabilized femtosecond pulses at multi-MHz repetition rates, in conjunction with advances in mirror coating technology enabling the construction of high-finesse broadband resonators. Today, the generation of radiation via nonlinear processes driven by femtosecond pulses, which are enhanced in passive resonators, exhibits the potential of approaching power levels necessary for applications which so far have only been demonstrated with synchrotron or free-electron laser radiation. In addition, the laser-based generation provides the prospect of unprecedented time and space coherence in the envisaged regions of the electromagnetic spectrum and exhibits the advantages of simplicity, compactness and low cost. Further development of the femtosecond enhancement cavity technology has motivated the research reported in this thesis.

1.2 Motivation: Enhancement of Nonlinear Processes in Passive Optical Resonators

An *enhancement cavity* (EC) is a passive optical resonator in which laser light is coherently overlapped. Coherence in this context refers to the condition that the input laser light has a constant phase relationship with the light "stored" in the EC, i.e. the intracavity light.[5]

Figure 1.2 shows a schematic of an EC. A train of input (also referred to as seeding) laser pulses, equidistant in time and with a fixed electric field phase relation, hits the input coupler of the EC. Under proper conditions, a high-power pulse forms in the cavity, circulating at the repetition rate of the initial pulse train (usually several

---

[5] This can be either satisfied by a single-frequency continuous-wave (CW) laser, or by a laser emitting several frequencies with a fixed phase relationship, which is the case of a mode-locked laser generating pulses. Since CW operation can be regarded as a special case of a mode-locked laser, the following discussion of pulsed EC's includes the CW case.

tens of MHz), and receiving energy from each of the seeding pulses.[6] The power enhancement in an EC is limited by the cavity losses and chromatic dispersion, and in state-of-the-art setups it amounts to a few orders of magnitude. EC's are ideally suited to boost the efficiency of nonlinear processes with low single-pass conversion efficiency (such as HHG or DFG) because firstly, the low conversion efficiency does not affect the enhancement considerably, enabling high intracavity intensities and secondly, the pulse used to drive the nonlinear process is recycled after each interaction with the nonlinear medium.[7]

EC's are often referred to as external resonators because they do not contain a gain medium. Driving nonlinear processes in such a resonator rather than in a laser cavity exhibits several additional advantages. The lack of elements prone to the high average and peak powers (such as the gain medium and/or the mode-locking components) circumvents their inherent limitations related to thermal and nonlinear effects and also facilitates operation under vacuum conditions, which is mandatory for certain applications. Moreover, the input laser light can be readily subjected to single-pass processes (such as amplification, nonlinear spectral broadening, pulse picking, pulse shaping etc.) before seeding the EC. Further advantages include the flexibility concerning the intracavity power regime which can be easily varied over a large range by adjusting the parameters of the seeding pulses, an outstanding intracavity beam quality and the action of the passive resonator as a high-frequency low-pass noise filter. The on-resonance sensitivity amplification of single-roundtrip dispersion and losses can also be used for accurate detection measurements.

The principle of resonant enhancement of an optical nonlinear process was demonstrated as early as in 1966 for SHG with a continuous-wave (CW) laser [7]. Further development followed, see [8–10] and references therein. Other EC's were constructed for pulsed SHG [11, 12], third harmonic generation of a CW laser [13], frequency mixing of two CW lasers [14], the generation of THz radiation via DFG [15] and recently, an EC-based optical parametric chirped-pulse amplification scheme was proposed [16]. First proofs of principle for EC-based HHG in a neutral noble gas jet have attracted high interest in recent years [17–23].[8] However, the power scaling limitations of these unique sources are still far from being reached.

At our institute there are two main motivations of pursuing further power scaling of high-repetition rate HHG in EC's. On the one hand, this technology offers the prospect of extreme nonlinear optics and in particular, isolated attosecond pulse generation at repetition rates of several tens of MHz, which is a few orders of magnitude larger than in current systems [24, 25]. This would open new perspectives in *attosecond physics*, allowing for new insights to both the collective and individual motion of electrons on atomic and molecular scales. On the other hand, efficient XUV generation with high repetition rates will enable high-precision *frequency comb*

---

[6] Note that multiple pulses can be stored in the EC in the same manner, if the cavity roundtrip time is a multiple of the pulse repetition period.

[7] The task of coupling out the new frequencies from the cavity strongly depends on the participating wavelengths.

[8] A more detailed overview of EC's for HHG is given in Appendix 7.1.

spectroscopy in a hitherto inaccessible spectral region [17]. In the frequency domain, the distance between the comb lines equals the repetition frequency of the ultrashort pulses and is transferred to the generated high harmonic spectrum. Fewer comb lines per unit of frequency implies more energy per comb line and the possibility of resolving single lines.

The significance of investigations related to the power scaling of the femtosecond enhancement cavity technique extends beyond these two particular applications. The availability of a table-top source of bright VUV and XUV radiation would undoubtedly result in a significant boost of pertinent experiments in numerous fields, including XUV microscopy and tomography [26], interferometry and holography [27], plasma physics [28], spectroscopy [29], at-wavelength-testing of components for EUV lithography [30] and surface and material studies [31], just to name the most prominent ones.[9] So far, some of these applications have only been demonstrated using synchrotron radiation. Besides HHG in a gas, the study and development of power scaling of the EC technology can be useful for arbitrary nonlinear processes. Therefore, the motivation of this work ties in with the basic preoccupation with the study of light, mentioned at the beginning of this chapter.

1.3 Overview of the Results

During the past four years, in the frame of this doctorate work a femtosecond enhancement cavity experiment was set up and investigations concerning the power scaling of this technology were performed. The results can be summed up into four categories:

1. Power scaling of the empty cavity. The EC presented in this thesis is seeded by an Yb-based laser system and supports 200-fs pulses with 20 kW of average power at a repetition rate of 78 MHz. Beyond this power level, intensity-related effects in the mirrors were identified, leading to resonator instabilities and damage. By extending the pulse duration to 2 ps by chirping the seed laser pulses, we could obtain 72 kW of intracavity circulating power with the maximum available input power of 50 W. To the best of our knowledge, this power level is the highest for ultrashort pulses at a multi-MHz repetition rate reported in the literature to this day and corresponds to an increase of roughly one order of magnitude with respect to the state of the art at the beginning of this work [32]. The limitations identified spawned the design of the next-generation EC in our group, following the strategy of increasing the spot size on the mirrors.

2. Advanced diagnostics. The first highly sensitive intracavity dispersion measurement method working at high intra-cavity intensities was developed, based on spatially and spectrally resolved interferometry (SSI) of a copy of the input beam to the cavity with a copy of the circulating beam. In particular, dispersion measurements of the empty cavity have confirmed the intensity-related power scaling limitations.

---

[9] A number of further exciting applications are summarized in Refs. [4, 5].

This method is expected to be instrumental to the optimization of the operation of EC's including nonlinear processes.

3. XUV output coupling. One of the main challenges of EC-based XUV generation is coupling out the intracavity generated high harmonic radiation from the EC, especially at the power levels demonstrated in our EC. As a part of this work, the suitability of several output coupling methods for high-power operation was investigated. Two novel methods were developed, offering the prospect of pushing the limits of high-power operation and efficient output coupling of the shortest wavelengths.

4. High-harmonic generation. First HHG experiments were carried out successfully at a moderate power level. With 200 fs pulse duration, 2.4 kW of average power and a focus radius $w_0 = 23\,\mu$ ($1/e^2$-intensity decay), harmonics up to order 17 generated in Xe were detected. The results are in excellent agreement with theoretical predictions. These experiments are intended as benchmarks for future scaling of the HHG process, especially employing one of the novel output coupling methods mentioned above, suited for the power regimes which have been demonstrated in the empty cavity.

In the following, an overview of the journal publications related to the results in these four categories, which I co-authored, is given:

1. Power scaling of the empty cavity. The main results are published in the following two papers, for which I performed the majority of the experimental work and manuscript preparation:

- I. Pupeza, T. Eidam, J. Rauschenberger, B. Bernhardt, A. Ozawa, E. Fill, A. Apolonski, Th. Udem, J. Limpert, Z. A. Alahmed, A. M. Azzeer, A. Tünnermann, T. W. Hänsch and F. Krausz, "Power scaling of a high repetition rate enhancement cavity," Opt. Letters **12**, 2052 (2010).
- I. Pupeza, T. Eidam, J. Kaster, B. Bernhardt, J. Rauschenberger, A. Ozawa, E. Fill, T. Udem, M. F. Kling, J. Limpert, Z. A. Alahmed, A. M. Azzeer, A. Tünnermann, T. W. Hänsch, and F. Krausz, "Power scaling of femtosecond enhancement cavities and high-power applications," Proc. SPIE **7914**, 791411 (2011).

In the frame of the collaboration with the Hänsch-group, I contributed (mainly with discussions) to the green-cavity project led by Birgitta Bernhardt. The following manuscript is currently in preparation:

- B. Bernhardt, A. Ozawa, A. Vernaleken, I. Pupeza, J. Kaster, Y. Kobayashi, R. Holzwarth, E. Fill, F. Krausz, T. W. Hänsch, and Th. Udem, "Ultraviolet Frequency Combs Generated by a Femtosecond Enhancement Cavity in the Visible," submitted for publication.

2. Advanced diagnostics. The dispersion measurement method is published in the following manuscript. While I contributed most of the experimental work, data interpretation and manuscript preparation, the initial idea of applying SSI as well as great support with the experiment and data interpretation came from Xun Gu, along with substantial support from the other authors:

- I. Pupeza, X. Gu, E. Fill, T. Eidam, J. Limpert, A. Tünnermann, F. Krausz, and T. Udem, "Highly Sensitive Dispersion Measurement of a High-Power Passive

Optical Resonator Using Spatial-Spectral Interferometry," Opt. Express **18**, 26184 (2010).

3. XUV output coupling. For the following paper I performed the majority of the experimental work and manuscript preparation:

- I. Pupeza, E. Fill, and F. Krausz, "Low-loss VIS/IR-XUV beam splitter for high-power applications," Opt. Express **19**, 12108 (2011).

For the following paper, I contributed mainly the experimental setup and the measurements described in Chap. 4, together with Jan Kaster, and helped with the manuscript preparation:

- Y.-Y. Yang, F. Süssmann, S. Zherebtsov, I. Pupeza, J. Kaster, D. Lehr, E.-B. Kley, E. Fill, X.-M. Duan, Z.-S. Zhao, F. Krausz, S. Stebbings, and M. F. Kling, "Optimization and characterization of a highly-efficient diffraction nanograting for MHz XUV pulses," Opt. Express **19**, 1955 (2011).

For the following two publications I performed the majority of the experimental work and contributed to the theory and manuscript preparation:

- J. Weitenberg, P. Russbüldt, T. Eidam, and I. Pupeza, "Transverse mode tailoring in a quasi-imaging high-finesse femtosecond enhancement cavity," Opt. Express **19**, 9551 (2011).
- J. Weitenberg, P. Russbüldt, I. Pupeza, T. Udem, H.-D. Hoffmann, and R. Poprawe, "Geometrical on-axis access to high-finesse resonators by quasi-imaging," manuscript in preparation.

4. High-harmonic generation. The HHG results reported in this thesis are preliminary and have not yet been published. However, they are very promising and a timely publication is being considered.

1.4 Structure of the Thesis

The subsequent chapters are organized as follows:

- Chapter 2 reviews the most relevant theoretical aspects of the area of this work, including the fundamentals of mode-locked lasers, ultrashort pulses, enhancement cavities and high-harmonic generation (HHG).
- Chapter 3 addresses the objectives of the experiments presented in this thesis and the technological challenges on the way to reaching them. An overview of the solutions worked out during this thesis and of the remaining challenges is given.
- Chapter 4 presents the experimental results elaborated during this work.
- Chapter 5 gives an outlook on future research motivated by and related to the results of this work.

- Chapter 6 resumes results obtained in the course of this doctorate work, which have been published in the journal articles [33–37]. These results are not repeated in Chaps. 1, 2, 3, 4 and 5 in detail. Rather, they are referred in this thesis with standard citations, followed by the remark "see also Chap. 6.

References

1. T. Walther, H. Walther, *Was ist Licht?* (C. H. Beck, München, 1999)
2. A. Siegman, *Lasers* (University Science Books, Sausalito, 1986)
3. F. Träger, *Springer Handbook of Lasers and Optics* (Springer, New York, 2007)
4. D. Attwood, *Soft X-rays and Extreme Ultraviolet Radiation* (Cambridge University Press, Cambridge, 1999)
5. P. Jaegle, *Coherent Sources of XUV Radiation* (Springer, New York, 2006)
6. M.J. Weber, Handbook of Laser Wavelengths (CRC Press, Boca Raton, 1999)
7. A. Ashkin, G.D. Boyd, J.M. Dziedzic, Resonant optical second harmonic generation and mixing. IEEE J. Quantum Electron. **2**, 109 (1966)
8. K. Fiedler, S. Schiller, R. Paschotta, P. Kürz, J. Mlynek, Highly efficient frequency-doubling with a doubly resonant monolithic total-internal-reflection ring resonator. Opt. Lett. **18**, 1786 (1993)
9. Z.Y. Ou, H.J. Kimble, Enhanced conversion efficiency for harmonic-generation with double-resonance. Opt. Lett. **18**, 1053 (1993)
10. R. Paschotta, P. Kürz, R. Henking, S. Schiller, J. Mlynek, 82% efficient continuous-wave frequency doubling of 1.06 μm with a monolithic MgO:LiNbO3 resonator. Opt. Lett. **19**, 1325 (1994)
11. V.P. Yanovsky, F.W. Wise, Frequency doubling of 100-fs pulses with 50% efficiency by use of a resonant enhancement cavity. Appl. Phys. Lett. **19**, 1952 (1994)
12. E. Peters, S.A. Diddams, P. Fendel, S. Reinhardt, T.W. Hänsch, T. Udem, A deep-UV optical frequency comb at 205 nm. Opt. Express **17**, 9183 (2009)
13. J. Mes, E.J. van Duijn, R. Zinkstock, S. Witte, W. Hogervorst, Third-harmonic generation of a continuous-wave Ti:Sapphire laser in asternal resonant cavities. Appl. Phys. Lett. **82**, 4423 (2003)
14. B. Couillaud, T.W. Hänsch, S.G. MacLean, High power CW sum-frequency generation near 243 nm using two intersecting enhancement cavities. Opt. Commun. **50**, 127 (1984)
15. M. Theuer, D. Molter, K. Maki, C. Otani, J.A. L'huillier, R. Beigang, Terahertz generation in an actively controlled femtosecond enhancement cavity. Appl. Phys. Lett. **93**, 041119 (2008)
16. F. Ilday, F.X. Kärtner, Cavity-enhanced optical parametric chirped-pulse amplification. Opt. Lett. **31**, 637 (2006)
17. C. Gohle, T. Udem, M. Herrmann, J. Rauschenberger, R. Holzwarth, H.A. Schuessler, F. Krausz, T.W. Hänsch, A frequency comb in the extreme ultraviolet. Nature **436**, 234 (2005)
18. R. Jones, K. D. Moll, M. J. Thorpe, J. Ye, Phase-Coherent Frequency Combs in the Vacuum Ultraviolet via High-Harmonic Generation inside a Femtosecond Enhancement Cavity. Phys. Rev. Lett. **94**, 193 201 (2005)
19. D.C. Yost, T.R. Schibli, J. Ye, Efficient output coupling of intracavity high harmonic generation. Opt. Lett. **33**, 1099–1101 (2008)
20. A. Ozawa, J. Rauschenberger, C. Gohle, M. Herrmann, D.R. Walker, V. Pervak, A. Fernandez, R. Graf, A. Apolonski, R. Holzwarth, F. Krausz, T. Hänsch, and T. Udem, High Harmonic Frequency Combs for High Resolution Spectroscopy. Phys. Rev. Lett. **100**, 253901 (2008)
21. A. Cingöz, D.C. Yost, J. Ye, A. Ruehl, M. Fermann, I. Hartl, Power scaling of high-repetition-rate HHG. International Conference on Ultrafast Phenomena, 2010

22. B. Bernhardt, A. Ozawa, I. Pupeza, A. Vernaleken, Y. Kobayashi, R. Holzwarth, E. Fill, F. Krausz, T.W. Hänsch, T. Udem, Green enhancement cavity for frequency comb generation in the extreme ultraviolet. CLEO, paper QTuF3, 2011
23. A. Ozawa, Y. Kobayashi, Intracavity high harmonic generation driven by Yb-fiber based MOPA system at 80 MHz repetition rate. CLEO, paper CThB4, 2011
24. T. Brabec, F. Krausz, Intense few-cycle laser fields: Frontiers of nonlinear optics. Rev. Mod. Phys. **72**, 545 (2000)
25. F. Krausz, M. Ivanov, Attosecond physics. Rev. Mod. Phys. **81**, 163 (2009)
26. A. Barty, S. Boutet, M. Bogan, S. Hau-Riege, S. Marchesini, K. Sokolowski-Tinten, N. Stojanovic, R. Tobey, H. Ehrke, A. Cavalleri, S. Düsterer, M. Frank, S. Bajt, B.W. Woods, M.M. Seibert, J. Hajdu, R. Treusch, H.N. Chapman, Ultrafast single-shot diffraction imaging of nanoscale dynamics. Nat. Photogr. **2**, 415 (2008)
27. P.W. Wachulak, M.C. Marconi, R.A. Bartels, C.S. Menoni, J.J. Rocca, Soft X-ray laser holography with wavelength resolution. Opt. Express **15**, 10622 (2007)
28. J. Filevich, K. Kanizay, M.C. Marconi, J.L.A. Chilla, J.J. Rocca, Dense plasma diagnostics with an amplitude-division soft-X-ray laser interferometer based on diffraction gratings. Opt. Lett. **25**, 356 (2000)
29. P.C. Hinnen, S.E. Werners, S. Stolte, W. Hogervorst, W. Ubachs, XUV-laser spectroscopy of HD at 92–98 nm. Phys. Rev. A **52**, 4425 (1995)
30. J. Lin, N. Weber, J. Maul, S. Hendel, K. Rott, M. Merkel, G. Schoenhense, U. Kleineberg, At-wavelength inspection of sub-40 nm defects in extreme ultraviolet lithography mask blank by photoemission electron microscopy. Opt. Lett. **32**, 1875 (2007)
31. A.L. Cavalieri, N. Müller, T. Uphues, V.S. Yakovlev, A. Baltuska, B. Horvath, B. Schmidt, L. Blümel, R. Holzwarth, S. Hendel, M. Drescher, U. Kleineberg, P.M. Echenique, R. Kienberger, F. Krausz, U. Heinzmann, Attosecond spectroscopy in condensed matter. Nature **449**, 1029 (2007)
32. I. Hartl, T.R. Schibli, A. Marcinkevicius, D.C. Yost, D.D. Hudson, M.E. Fermann, J. Ye, Cavity-enhanced similariton Yb-fiber laser frequency comb: 3×10^{14} W/cm^2 peak intensity at 136MHz. Opt. Lett. **32**, 2870 (2007)
33. I. Pupeza, T. Eidam, J. Rauschenberger, B. Bernhardt, A. Ozawa, E. Fill, A. Apolonski, T. Udem, J. Limpert, Z.A. Alahmed, A.M. Azzeer, A. Tünnermann, T.W. Hänsch, F. Krausz, Power scaling of a high repetition rate enhancement cavity. Opt. Lett. **12**, 2052 (2010)
34. I. Pupeza, X. Gu, E. Fill, T. Eidam, J. Limpert, A.Tünnermann, F. Krausz, T. Udem, Highly sensitive dispersion measurement of a high-power passive optical resonator using spatial-spectral interferometry. Opt. Express **18**, 26184 (2010)
35. I. Pupeza, E. Fill, F. Krausz, Low-loss VIS/IR-XUV beam splitter for high-power applications. Opt. Express **19**, 12108 (2011)
36. Y.-Y. Yang, F. Süssmann, S. Zherebtsov, I. Pupeza, J. Kaster, D. Lehr, E.-B. Kley, E. Fill, X.-M. Duan, Z.-S. Zhao, F. Krausz, S. Stebbings, M.F. Kling, Optimization and characterization of a highly-efficient diffraction nanograting for MHz XUV pulses. Opt. Express **19**, 1955 (2011)
37. J. Weitenberg, P. Russbüldt, T. Eidam, I. Pupeza, Transverse mode tailoring in a quasi-imaging high-finesse femtosecond enhancement cavity. Opt. Express **19**, 9551 (2011)

Chapter 2
Theoretical Background

2.1 Passive Enhancement of Ultrashort Pulses

The setup of any femtosecond enhancement cavity (EC) experiment consists of three main components. Firstly, the pulses are generated by a mode-locked oscillator and optionally post-processed (e.g. amplified and/or spectrally broadened and temporally compressed etc.). The second component is the EC in which the laser pulses are overlapped coherently. If the EC is used for nonlinear conversion, it also contains the respective mechanism. And thirdly, one or more feedback loops are needed to ensure an interferometric overlap of the seeding laser pulses with the pulse(s) circulating in the enhancement cavity.

Ideally, a steady state is aimed for in which the enhanced electric field circulating in the passive cavity is a power scaled version of the seeding laser field. However, due to the fact that the spectrum emitted by a mode-locked laser consists of equidistant modes and the resonances of a passive cavity are in general not equidistant in the frequency domain, this task is highly challenging. To explain this challenge, we provide a theoretical background of the three main components. The different nature of the mechanisms underlying the generation of the pulses and their passive enhancement is addressed, and aspects particularly relevant for the experimental results presented in this thesis are emphasized.

2.1.1 Ultrashort Pulses from Mode-Locked Lasers

Mode-locking underlies the generation of the shortest light pulses directly from a laser oscillator. The advent of self-mode-locking techniques (see. e.g. [1]) merely 20 years ago and the subsequent development of phase stabilization in the last decade have had a significant impact on scientific research and technology. With this technique, light pulses comprising only a few oscillations of the electric field can be generated, confining the available laser energy to pulse durations on a femtosecond timescale.

I. Pupeza, *Power Scaling of Enhancement Cavities for Nonlinear Optics*,
Springer Theses, DOI: 10.1007/978-1-4614-4100-7_2,
© Springer Science+Business Media New York 2012

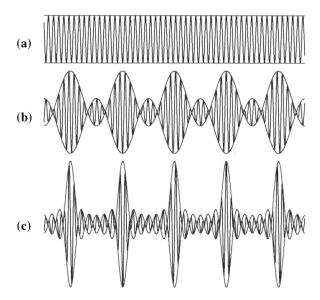

Fig. 2.1 The formation of ultrashort pulses: superposition of axial modes of an ideal Fabry-Perot resonator, oscillating with the same phase. The *red line* indicates the intensity profile for **a** a single mode, **b** for 3 and **c** for 7 modes

In addition, phase stabilization guarantees the pulse-to-pulse reproducibility of the electric field within the pulse intensity envelope. Pulses with these characteristics have marked the birth of ultrafast and attosecond science [1, 2], studying and controlling light-matter interactions at unprecedented intensity levels and atomic dynamics on previously inaccessible time scales. Phase-stabilized mode-locked lasers have also enabled high-precision optical frequency metrology. Their frequency spectrum is a regular structure of equidistant lines, called *frequency comb*, see e.g. [3–8]. The frequency comb can be parameterized by using only two radio frequencies (RF), which can be easily measured, e.g. with respect to the hyperfine transition of cesium at 9.193 GHz which provides the standard for one second. This constitutes a bridge transferring the measurement precision over the frequency gap between the RF spectrum and optical frequencies. A textbook-level description of the mode of operation of mode-locked lasers can be found e.g. in [9] or [10]. Here, we restrict the description to an overview of the basic mechanisms leading to the formation of a train of femtosecond pulses with a fixed phase relationship.

Mode locking refers to imposing a certain, fixed phase relation to the resonant axial (longitudinal) modes of a laser cavity. Figure 2.1 visualizes the intensity of (a) a single mode, i.e. continuous-wave operation, and the superposition of (b) 3 and (c) 7 modes of an ideal Fabry-Perot resonator, if these modes oscillate with identical phases. At a fixed time, or at a fixed space, the modes interfere constructively at equidistant maxima in space, or in time, respectively, which leads to the formation of light pulses. The more modes simultaneously resonant in the cavity, i.e. the broader the bandwidth of contributing frequencies, the shorter the resulting pulses. However, in a real laser cavity, the circulating light pulses are subjected to dispersion, i.e. to a nonlinear dependence of the wavenumber $k(\omega)$ on the angular frequency ω. In general, for a pulse with a spectrum centered around the frequency ω_c, the wavenumber $k(\omega)$ can be expanded in a Taylor series:

$$k(\omega) = \frac{\omega_c}{v_{ph}} + \frac{\omega - \omega_c}{v_{gr}} + \frac{1}{2}GVD(\omega - \omega_c)^2 + \ldots, \tag{2.1}$$

where v_{ph}, v_{gr} and GVD denote the phase velocity of propagation of the frequency component ω_c, the group velocity of the pulse and the group velocity dispersion, respectively (cf. e.g. Sect. 9.1 in [11]). The terms of orders 2 (i.e. GVD) and higher describe the dispersion accumulated by the pulse upon propagation and thus, the pulse lengthening in time. In the presence of non-zero net roundtrip dispersion, a pulse circulating in the laser cavity would broaden infinitely in time. In a mode-locked laser, however, the pulse reproduces after each round trip, which is equivalent to the fact that the effect of residual cavity dispersion on the pulse is compensated for upon each roundtrip. The intensity evolution of the pulse is given by the relative phase of the frequency components. Therefore, the nonlinear process of self phase modulation (SPM) can be used to perform this compensation and acts as a simultane-ous switch for all active modes. The effect of SPM upon a roundtrip is determined by the interplay of the intracavity nonlinear mechanisms, such as the amplification and propagation in the active medium and (usually) an additional mode-locking mech-anism. Dimensioning the phase effects in a mode-locked oscillator can be done by solving the nonlinear Schrödinger equation, see e.g. Sect. 10.3 in [11]. The interplay between the cavity dispersion, the mode-locking mechanism and the active material net gain bandwidth also determines the (achievable) output pulse duration of a mode-locked laser. A mathematical model of the mode-locking process can be found e.g. in [12, 13]. A decisive result is that the mode locking mechanism counterbalances the terms of order 2 and higher in the Taylor expansion in Eq. (2.1), implying:

$$k(\omega) = \frac{\omega_c}{v_{ph}} + \frac{\omega - \omega_c}{v_{gr}}. \tag{2.2}$$

This means in particular that the pulse can propagate with a different group ve-locity v_{gr} than the phase velocity v_{ph} of the wave with frequency ω_c. The following derivation will show that this fact is of crucial importance for the control of the electric field of the generated pulse train.

The steady-state resonance boundary condition of a linear Fabry-Perot oscillator with length L (or $2L$ for a ring cavity) is given by:

$$2k(\omega_N)L = 2\pi N, \tag{2.3}$$

where the integer N denotes the axial (longitudinal) mode number. The last two equations determine the frequencies of the modes of a mode-locked oscillator:

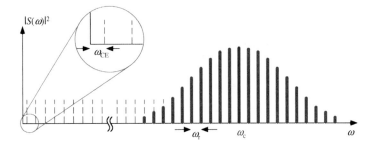

Fig. 2.2 Frequency-domain representation of the frequency comb emitted by a phase-stabilized mode-locked laser (as in [5]). The comb parameters are: the repetition frequency ω_r and the carrier-envelope frequency ω_{CE}

$$\omega_N = N\frac{2\pi}{2L}v_{gr} + \omega_c\left(1 - \frac{v_{gr}}{v_{ph}}\right). \tag{2.4}$$

Thus, the frequency difference between two adjacent modes amounts to:

$$\omega_{N+1} - \omega_N = \frac{2\pi}{2L}v_{gr} =: \omega_r \tag{2.5}$$

and corresponds to 2π times the repetition frequency of the pulses, which we will denote by ω_r.

The second term on the right-hand side of Eq. (2.4) represents an offset frequency, which we will denote by ω_{CE}:

$$\omega_{CE} := \omega_c\left(1 - \frac{v_{gr}}{v_{ph}}\right). \tag{2.6}$$

The abbreviation CE stands for *carrier-envelope* due to the significance of ω_{CE} in the time domain, which we will address shortly. In conclusion, the spectrum emitted by a stable mode-locked laser is a *comb of equidistant modes* with a spacing ω_r and offset from a multiple of ω_r by ω_{CE}:

$$\omega_N = N\omega_r + \omega_{CE}. \tag{2.7}$$

Figure 2.2 shows such a spectrum. With a spectral amplitude envelope $S(\omega)$, the electric field of the pulses in the time domain can be written as a Fourier series:

$$E(t) = \sum_{N=-\infty}^{\infty} S(\omega_N)e^{-j(N\omega_r + \omega_{CE})t}. \tag{2.8}$$

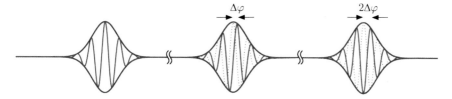

Fig. 2.3 Pulse-to-pulse evolution of the carrier-envelope offset for $\Delta\varphi = \pi/2$ (as in [5])

One cavity roundtrip period $T = 2\pi/\omega_r$ later, the electric field equals:

$$E(t+T) = \sum_{N=-\infty}^{\infty} S(\omega_N)e^{-j(N\omega_r+\omega_{CE})(t+T)}$$

$$= E(t)e^{-j(2N\pi+2\pi\omega_{CE}/\omega_r)}$$

$$= E(t)e^{-j\Delta\varphi}, \text{ with:} \tag{2.9}$$

$$\Delta\varphi := 2\pi\frac{\omega_{CE}}{\omega_r}. \tag{2.10}$$

In other words, the electric field of each pulse produced by the laser is a copy of the electric field of the previous pulse, shifted by a phase $\Delta\varphi$. Due to the often used time-domain representation of the electric field as a continuously oscillating carrier wave times an envelope function, $\Delta\varphi$ is also called the "carrier-envelope offset phase slippage", which also explains the name of ω_{CE}. Figure 2.3 visualizes the evolution of the carrier-envelope offset for the slippage phase $\Delta\varphi = \pi/2$. Stabilizing the pulse-to-pulse phase of a mode-locked laser is crucial for localizing the comb modes in the frequency domain as well as for time-domain applications which demand the reproducibility of the electric field rather than just that of the intensity profile. Carrier-envelope phase measurement and stabilization has been made possible in the last decade, see e.g. [3–8, 14–16].

The basic property enabling the enhancement of the pulses generated by a mode-locked oscillator in a passive cavity is the pulse-to-pulse reproducibility of the electric field, which is equivalent to the comb spectrum. Depending on the purpose of the enhancement, the pulses generated by the mode-locked oscillator can be further processed without affecting this basic property. Such processes include: *pulse picking:* reduction of the repetition rate; this generates new spectral components, in such a way that the comb spacing equals the reduced repetition rate; *sideband generation:* generation of additional frequency components through fast modulation, particularly useful for the Pound-Drever-Hall locking scheme (see Sect. 2.1.3); *pulse chirping:* manipulation of the spectral phase, in particular, allows for a variation of the pulse peak power while keeping the spectral components and the energy constant; *linear power amplification:* scaling up the pulse energy while keeping the other parameters constant; *comb-preserving harmonic generation:* creates a frequency comb with the original comb spacing at a multiple of the central frequency.

2.1.2 Passive Enhancement in an External Cavity

Similar to a mode-locked laser, the steady-state condition of an enhancement cavity requires a pulse which is reproduced after each roundtrip. However, in contrast to a mode-locked oscillator, the latter does not incorporate an intracavity mechanism for dispersion compensation. Rather, the optical resonances of cavity, which are continuously excited by the seeding frequency comb modes, are not equidistant in the frequency domain due to roundtrip dispersion. This distinction from the seeding frequency comb affects the interference of the seeding with the circulating field, representing an enhancement limitation. Moreover, the fact that the circulating field has a different spectral amplitude and phase distribution than the seeding field requires increased attention when controlling the interference of the two. In the following, we first derive an analytical model for the steady state of a passive cavity excited by a frequency comb. Then, we use this model to discuss the trade-off between enhancement finesse and bandwidth.

The Steady State

Figure 2.4 shows the complex electric fields at the input coupler (IC) of an enhancement cavity in the steady state. Usually, enhancement cavities for nonlinear conversion are implemented as ring cavities. This is to avoid a double pass in opposite directions through the nonlinear material and to enable a straightforward spatial separation of the field seeding the cavity and the field reflected by the IC.[1] The following derivation, however, holds for both linear and ring cavities. Moreover, we assume perfect transverse mode matching at the IC, i.e. that the seeding laser beam is matched to the excited cavity transverse mode. For each frequency component ω, the single-roundtrip power attenuation and accumulated phase are denoted by $A(\omega)$ and $\theta(\omega)$, respectively. Thus, one cavity roundtrip of the circulating electric field component $\widetilde{E}_c(\omega)$ can be completely described by its multiplication by $\sqrt{A(\omega)} \exp[j\theta(\omega)]$. Let $R(\omega)$ denote the IC power reflectivity, so that a reflection on the cavity side at the IC implies the multiplication of the impinging electric field $\widetilde{E}_c(\omega)$ by $\sqrt{R(\omega)}$. Let $T(\omega)$ denote the IC power transmission. Resonant enhancement requires the constructive interference of the intracavity field reflected by the IC, i.e., $\sqrt{R(\omega)}\widetilde{E}_c(\omega)\sqrt{A(\omega)}\exp[j\theta(\omega)]$ with the portion of the input field transmitted through the IC, i.e., $\sqrt{T(\omega)}\widetilde{E}_i(\omega)$. In the steady state the sum of these two interfering fields equals $\widetilde{E}_c(\omega)$:

---

[1] This comes in handy especially when using seeding systems sensitive to optical feedback.

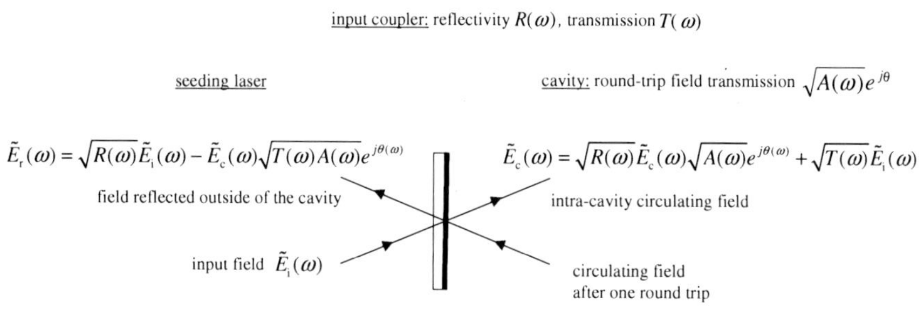

input coupler: reflectivity $R(\omega)$, transmission $T(\omega)$

seeding laser

cavity: round-trip field transmission $\sqrt{A(\omega)}e^{j\theta}$

$\tilde{E}_r(\omega) = \sqrt{R(\omega)}\tilde{E}_i(\omega) - \tilde{E}_c(\omega)\sqrt{T(\omega)A(\omega)}e^{j\theta(\omega)}$

field reflected outside of the cavity

$\tilde{E}_c(\omega) = \sqrt{R(\omega)}\tilde{E}_c(\omega)\sqrt{A(\omega)}e^{j\theta(\omega)} + \sqrt{T(\omega)}\tilde{E}_i(\omega)$

intra-cavity circulating field

input field $\tilde{E}_i(\omega)$

circulating field
after one round trip

Fig. 2.4 Electric fields at the input coupler of an enhancement cavity in the steady state

$$\widetilde{E_c}(\omega) = \sqrt{R(\omega)}\widetilde{E_c}(\omega)\sqrt{A(\omega)}\exp[j\theta(\omega)] + \sqrt{T(\omega)}\tilde{E}_i(\omega) \qquad (2.11)$$

$$\Leftrightarrow \tilde{H}(\omega) := \frac{\widetilde{E_c}(\omega)}{\widetilde{E_i}(\omega)} = \frac{\sqrt{T(\omega)}}{1 - \sqrt{R(\omega)A(\omega)}\exp[j\theta(\omega)]}. \qquad (2.12)$$

For linear, i.e. intensity-independent behavior of the cavity, the Airy function $\tilde{H}(\omega)$ defined in Eq. (2.12) represents the transfer function of the cavity. However, the ratio $\tilde{H}(\omega)$ evaluated over the spectrum of the input field also bears significance if intracavity nonlinear processes are involved. In particular, $\tilde{H}(\omega)$ can be measured with high accuracy (see Sect. 4.2.2) providing information on the nonlinear process itself.

In a similar fashion, the superposition $\tilde{E}_r(\omega)$ of the input field reflected by the IC outside of the cavity and the intracavity field transmitted through the IC can be calculated:

$$\tilde{E}_r(\omega) = \sqrt{R(\omega)}\tilde{E}_i(\omega) - \widetilde{E_c}(\omega)\sqrt{T(\omega)A(\omega)}\exp[j\theta(\omega)]. \qquad (2.13)$$

Note that the minus sign between the two terms stems from a phase shift of opposed sign compared to the one between the two terms on the right-hand side of Eq. (2.11). This follows from the Fresnel equations and can also be explained by energy conservation. By plugging Eq. (2.12) in the above equation, we obtain:

$$\frac{\widetilde{E_r}(\omega)}{\widetilde{E_i}(\omega)} = \frac{\sqrt{R(\omega)} - [R(\omega) + T(\omega)]\sqrt{A(\omega)}\exp[j\theta(\omega)]}{1 - \sqrt{R(\omega)A(\omega)}\exp[j\theta(\omega)]}. \qquad (2.14)$$

The entire seeding energy is coupled to the cavity if the numerator of the right-hand side of Eq. (2.14) equals 0. This *impedance matching* condition is fulfilled for a frequency ω, if on the one hand the roundtrip phase at this frequency is a multiple of 2π and on the other hand the reflectivity and transmission of the IC are matched to the roundtrip amplitude losses. For a lossless IC, i.e. if $R(\omega) + T(\omega) = 1$ holds, the second impedance matching condition becomes $R(\omega) = A(\omega)$.

Two further notions, historically stemming from optical interferometry, are important in the context of passive cavities. The *free spectral range FSR* of a resonator (or interferometer) denotes the resonance spacing in the frequency domain. For a dispersive cavity, the *FSR* is ω-dependent and $FSR(\omega) = \pi c_0/[2n(\omega)L]$ holds, where c_0, $n(\omega)$ and $2L$ denote the speed of light in vacuum, the frequency-dependent refractive index for the intracavity propagation and the cavity geometrical length, respectively. The cavity *finesse* $\mathcal{F}$ is a measure of the resolving power of the resonator used as a transmission filter (in the sense of an etalon or interferometer). It is defined as the ratio of the *FSR* to the FWHM bandwidth of a cavity resonance $\Delta\omega_{cav}$, see e.g. Sect. 11.5 in [11]:

$$\mathcal{F}(\omega) := \frac{FSR(\omega)}{\Delta\omega_{cav}}. \tag{2.15}$$

With the field propagation coefficient $r_{tot}(\omega) := \sqrt{R(\omega)A(\omega)}$ summing up the total losses of the frequency ω upon a roundtrip propagation in the passive resonator, the following equation holds for $\Delta\omega_{cav}$ (cf. Eq. (52) in [11], Sect. 11.5):

$$\Delta\omega_{cav} = \frac{4c}{2L} \arcsin\left(\frac{1 - r_{tot}(\omega)}{2\sqrt{r_{tot}(\omega)}}\right) \tag{2.16}$$

$$\approx \left(\frac{1 - r_{tot}(\omega)}{\pi\sqrt{r_{tot}(\omega)}}\right) \cdot FSR(\omega). \tag{2.17}$$

It follows for the finesse:

$$\mathcal{F}(\omega) \approx \frac{\pi\sqrt{r_{tot}(\omega)}}{1 - r_{tot}(\omega)} \tag{2.18}$$

$$\approx \frac{2\pi}{1 - r_{tot}^2(\omega)} = \frac{2\pi}{1 - R(\omega)A(\omega)}. \tag{2.19}$$

Thus, the finesse is given by the resonator losses and is independent of the resonator length.

Summary of Useful Formulas

In the following, we sum up the main equations for two important special cases:

- *Assumptions.* No dispersion, i.e. ω-dependence is discarded, lossless IC, i.e. $R + T = 1$ holds, and the cavity is on resonance. Then Eqs. (2.12) and (2.14) imply for the power enhancement P and the reflected portion of the input light $Refl$:

$$P := |\widetilde{H}(\omega)|^2 = \left|\frac{\widetilde{E_c}(\omega)}{\widetilde{E_i}(\omega)}\right|^2 = \frac{T}{\left(1 - \sqrt{RA}\right)^2}, \tag{2.20}$$

$$Refl := \left| \frac{\widetilde{E}_r(\omega)}{\widetilde{E}_i(\omega)} \right|^2 = \left(\frac{\sqrt{R} - \sqrt{A}}{1 - \sqrt{RA}} \right)^2 . \qquad (2.21)$$

• *Assumptions.* Like above but also impedance matching is given, i.e. $R = A$ holds:

$$P = \frac{1}{T}, \qquad (2.22)$$

$$Refl = 0, \qquad (2.23)$$

$$\mathcal{F} = \pi P. \qquad (2.24)$$

Enhancement Trade-off: Finesse Versus Bandwidth

The roundtrip phase $\theta(\omega)$ describes the phase changes of a wave of frequency ω upon propagating in the cavity (usually in a transverse eigen-mode of the cavity) and interacting with its optics. Let us assume that a certain frequency ω_c is resonant in the cavity, i.e. the roundtrip phase accumulated by a wave with this frequency is a multiple of 2π. Then, $\theta(\omega)$ can be expanded in a Taylor series about ω_c:

$$\theta(\omega) = \theta_0 + GD(\omega - \omega_c) + GDD(\omega - \omega_c)^2 + \dots, \qquad (2.25)$$

where θ_0, GD and GDD denote a constant phase, the group delay and the group delay dispersion,[2] respectively. The constant phase term θ_0 has only a physical significance if a phase difference is considered, e.g. between two transverse modes of different orders (cf. Sect. 2.1.4). The group delay is determined by the repetition frequency of the pulse circulating in the cavity according to $GD = 2\pi/\omega_r$ and can usually be easily varied by tuning the geometrical cavity length.

Figure 2.5 shows the magnitude and phase of $\widetilde{H}(\omega)$ for an impedance-matched (dispersion-free) cavity with a power enhancement of 100, which is resonant at ω_c, over a frequency range extending 5% of ω_r to the left and to the right of ω_c. If a continuous wave with a slightly different frequency than ω_c is exciting this cavity resonance (see e.g. green line in Fig. 2.5), the wave oscillating in the resonator will not only experience a smaller power enhancement than that of a wave with ω_c, but also a constant phase shift with respect to the exciting wave.

Equation (2.12) implies that roundtrip dispersion, i.e. GDD and higher-order terms in Eq. (2.25) and also frequency-dependencies of $\sqrt{A(\omega)}$ affect both the amplitude and the phase of $\widetilde{H}(\omega)$. In particular, this means that the resonances of a real cavity are not equidistant in contrast to the seeding frequency comb. Thus, some of the

---

[2] Note that the GDD and the GVD are linked by the relationship $\theta(\omega) = k(\omega)d$ which implies $GDD(\omega) = GVD(\omega)d$, where d is a distance. While the GVD measures the dispersion upon propagation through a material in general, the GDD is used to characterize the dispersion of optical elements with a fixed geometry.

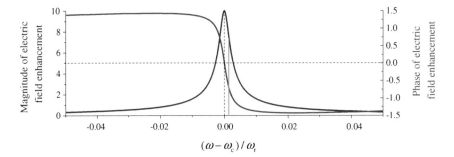

Fig. 2.5 Resonance of an impedance-matched cavity with a power enhancement of 100, centered at ω_c. *Green line* continuous wave with a slightly different frequency than ω_c, would experience a lower enhancement and a phase shift in the cavity

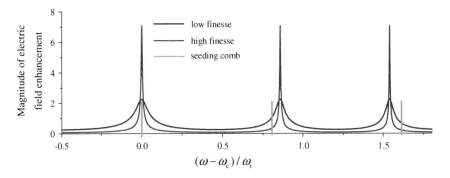

Fig. 2.6 Qualitative illustration of the enhancement finesse versus bandwidth trade-off. *Green* seeding comb frequencies, *red* high-finesse cavity with dispersion, *blue* low-finesse cavity with same dispersion. Spectral filtering and phase distortions are less pronounced for the low-finesse cavity. However, the power enhancement at the central mode is also smaller than in the high-finesse case

comb modes will necessarily be offset from the centers (magnitude peaks) of the corresponding cavity resonances, having a twofold effect on the circulating field. Firstly, the spectrum will not be evenly enhanced, meaning a spectral filtering of the seeding frequencies. Secondly, the phase of the circulating field will differ from the one of the input field. The dispersion imposed on the circulating field amounts to approximately the power enhancement factor times the roundtrip dispersion of the cavity (a derivation hereof can be found in Sect. 2.4 of [17]). A higher cavity finesse increases both effects. Thus, when dimensioning an enhancement cavity, a trade-off between finesse and bandwidth is necessary. Figure 2.6 illustrates this trade-off. The same frequency comb (green lines) is enhanced in a cavity with high finesse and in a cavity with low finesse. The high-finesse cavity enhances the central mode to a high degree but filters out the other modes strongly and also strongly affects the phase of the circulating field. The spectral filtering and the applied dispersion are much smaller in the case of the low-finesse cavity. However, the power enhancement in the latter is also smaller. This trade-off is clearly noticeable in the enhancement cavity systems demonstrated so far, cf. Appendix 7.1.

2.1.3 Interferometric Stabilization

The strict condition of interferometric overlap of the input field with the intracavity circulating field at all times, usually implies the need for active locking of the seeding comb to the cavity resonances or viceversa. The example in Appendix 7.2 shows that even small length deviations on a picometer scale, common to a normal lab environment, can alter the enhancement significantly. In this section we first address the control variables in the context of comb-cavity locking and review the most common locking schemes. Then, we derive quantitative conditions under which the lock of the seeding comb and the enhancement cavity can be realized with a single feedback loop. This is particularly relevant for the design of an experimental setup.

In our discussion we assume (i) that all fluctuations are slow enough and (ii) that the cavity response $\widetilde{H}(\omega)$ is linear, so that the steady states described in the previous two sections can be assumed. These assumptions usually apply for most mechanical and electronic distortions and they allow the discussion to be carried out in the frequency domain by using the comb and the cavity models given by Eqs. (2.7) and (2.12), respectively. We stress that processes on shorter time scales, such as nonlinear intracavity interactions, might introduce additional fluctuations, which are not accounted for by this discussion. Sect. 3.2.2) addresses this in more detail.

The Optimum Overlap

The frequency comb described by Eq. (2.7) can be parameterized by using two independent parameters. The two parameters need not necessarily be ω_r and ω_{CE} and for the sake of generality we will call them $p_{laser,1}$ and $p_{laser,2}$. In contrast, under the assumptions made above, the comb-*like*[3] structure of the enhancement cavity response in the frequency domain, given by Eq. (2.12), is completely described by the propagation length along the cavity optical path for a given set of (dispersive) cavity optics. In other words, the cavity transfer function can be parameterized by a single parameter (once the cavity optics are given), which we will refer to as p_{cavity}.

In analogy to the trade-off between finesse and enhancement bandwidth discussed in the previous section, the spectral filtering property of the cavity owed to dispersion implies that different combinations of $p_{laser,1}$, $p_{laser,2}$ and p_{cavity} might be optimal for different enhancement purposes. For instance, obtaining the shortest possible circulating pulse or obtaining the highest enhancement in a certain spectral region might call for a different overlap between the frequency comb modes and the cavity

---

[3] By the formulation "comb-*like*" we wish to emphasize that the cavity resonances are not equidistant.

Fig. 2.7 Feedback-loop connection of the parameters
$p_{laser,1}$, $p_{laser,2}$ and p_{cavity},
enabling optimum overlap
of the seeding comb with
the cavity resonances. Active
locking is indicated by the
arrows

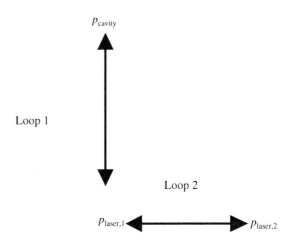

resonances in the same setup (cf. e.g. [18]). In general, there exists an *optimum over-lap* of the seeding comb modes with the cavity resonances, which can be described by a well defined combination of $p_{laser,1}$, $p_{laser,2}$ and p_{cavity}.

The train of thought followed so far has two immediate consequences. On the one hand, the single cavity parameter does not in general suffice to reach optimum overlap, since the seeding comb requires two parameters for a full description. Thus, at least one degree of freedom of the overlap needs to be controlled by means of the seeding comb. On the other hand, if one of the comb parameters, say $p_{laser,1}$, and the cavity parameter p_{cavity} are locked, the other comb parameter, i.e. $p_{laser,2}$, is unambiguously determined by $p_{laser,1}$ according to a constraint set by the optimum overlap (e.g. shortest intracavity pulse or highest intracavity power). In particular, this means that $p_{laser,2}$ can be locked to $p_{laser,1}$, without a direct feedback from the cavity, which is illustrated in Fig. 2.7.

Choice of the Control Parameters: Practical Considerations

The main fluctuations causing deviations from the optimum overlap and occurring in every real experimental setup are of mechanical and electrical nature. In general, mechanical vibrations affect all parameters, while electrical fluctuations affect mostly the laser (see e.g. [19, 20]), since the cavity is passive. Due to the effect of mechanical vibrations on the lengths of both the oscillator cavity and the enhancement cavity, it is common to stabilize the length of one of these cavities to match the length of the other one. If an enhanced frequency comb is desired for which the absolute frequencies are highly precise (e.g. for spectroscopy), then it is customary to lock the enhancement cavity to the seeding laser (which is locked to an external reference). However, if the precise knowledge of the absolute value of the enhanced modes is not mandatory, as is the case with most time-domain applications, the passive cavity can be used as a reference and one of the oscillator parameters can be locked to the cavity parameter. In practice this implementation is usually less demanding due to the lower power in the oscillator cavity (see also Sect. 4.1). In conclusion, we can thus assume without loss of generality that the cavity resonances are fixed, i.e. p_{cavity} is constant, and the two comb parameters are used to compensate deviations from the optimum overlap, which can be divided into two independent locking tasks (cf. Fig. 2.7):

- Loop 1: $p_{laser,1}$ controls the frequency of a comb mode (e.g. ω_c), so that this mode is locked to (usually the peak of) a cavity resonance, which is determined by p_{cavity}.
- Loop 2: $p_{laser,2}$ is set by using only feedback from $p_{laser,1}$, with the boundary condition that optimum overlap is reached.

The complex dynamics of mode locking (see e.g. [13, 19]) makes it very difficult to vary the two comb parameters ω_r and ω_{CE} completely independently. In general, the available control mechanisms (like piezoelectric actuators, acousto- and electro-optical modulators, dispersion management, optical pump power variation, etc.) act on both ω_r and ω_{CE}. Therefore, in practice, the parameters $p_{laser,1}$ and $p_{laser,2}$ are two linear combinations of ω_r and ω_{CE}, which need to satisfy two necessary conditions: (i) they need to be linearly independent and (ii) they need to be tunable over a region large enough around the optimum overlap so that all occurring fluctuations can be compensated for. Under these two conditions, the comb needs to be brought in the vicinity of the optimum overlap in the parameter space defined by $p_{laser,1}$ and $p_{laser,2}$ by coarse adjustment and, subsequently, active stabilization will ensure an optimum overlap. In the following paragraph, we review the most common schemes for the two locking tasks, i.e. $p_{laser,1}$ to p_{cavity} and $p_{laser,2}$ to $p_{laser,1}$.

Common Locking Schemes

An active feedback loop consists of three main components: an error signal, indicating the deviation of the actuating variable from the desired optimum, locking electronics setting the control mechanism according to this error signal, and the control mechanism itself. For the lock of a comb line to a Fabry-Perot cavity with an Airy-function response as given by Eq. (2.12) and plotted in Fig. 2.5, the control mechanism can be a piezo-actuated mirror, the optical pump power of the oscillator, an acousto- or electro-optical modulator etc.. The locking electronics are usually a standard, commercially available phase-locked loop. Rather than providing a deep insight into the mode of operation of these standard electronic components, we address in the following, the generation of the error signal, which represents the interface of the optical setup, and the task of interferometric stabilization, which is an electronic one.

The necessary property every error signal needs to fulfill, is bipolarity, i.e. the signal should indicate unambiguously the direction (and magnitude) of the drift which needs to be compensated for. For example, the intensity of the optical signal transmitted through a cavity mirror, as a function of the mismatch between the driving frequency and the cavity resonance, is not suited as an error signal for locking the peak of the resonance, since the Airy function is an even function (see Fig. 2.5). However, this signal is suited to lock the driving frequency to a sub-maximum level on one side of the Airy fringe, e.g. the position indicated by the green comb line in Fig. 2.5. This scheme is usually referred to as *side-of-fringe locking* and was first employed to stabilize (CW) lasers to a reference cavity [21] rather than to enhance a field with the purpose of increasing the available power.

Schemes able to lock a driving frequency to the peak of a cavity resonance require a bipolar error signal that equals 0 at the peak of the resonance and changes its sign when the mismatch passes through this zero. Such signals are usually obtained from observing the interference of a portion of light in phase with the driving frequency ("reference") and a portion of intracavity light that contains the phase information of the cavity ("sample").

The *Hänsch-Couillaud* or *polarization locking* scheme [22] works for polarization discriminating cavities. The linearly polarized light seeding the cavity needs to contain nonzero components in two orthogonal polarization directions, e.g. parallel and perpendicular to the intracavity polarization discrimination direction. For each polarization component, the overall field reflected by the cavity is given by Eq. (2.14) and represents the superposition of a portion of the seeding light which serves as the "reference" and a portion of transmitted light, which carries the phase information of the cavity and serves as the "sample" part of the interferometer. The "sample" portion is linearly polarized, but, due to the polarization discrimination, its decomposition along the two polarization directions has a different ratio of coefficients than the reference part of the beam. When the cavity is on resonance, the phase difference between the two linearly polarized parts is 0 so that their superposition is linearly polarized. However, when the cavity is off-resonance, the sample part has a frequency-dependent phase shift with respect to the reference part, manifesting itself as an elliptical polarization of the reflected beam. The direction and the magnitude of the ellipticity can readily be detected with an analyzer consisting of a quarter-wave plate, a polarizing beam splitter and a difference photodiode, providing the error signal. The cavity in the proof-of-principle paper [22] contains a Brewster plate which provides strong polarization discrimination. However, recently we have shown that the nonorthogonal incidence of the beam on the cavity mirrors provides a polarization discrimination large enough for a stably working Hänsch-Couillaud lock [17, 23]. The results presented in this thesis have been obtained using the Hänsch-Couillaud locking scheme.

Several locking schemes, which we summarize under the name of *transverse mode mismatch locking*, use the fact that an optically stable cavity has a set of well-defined transverse eigen-modes. If this cavity is excited externally and a mismatch between the exciting beam and the excited transverse mode is introduced, a spatial interference of the part of the input beam which is not matched to the cavity mode (reference) with the beam resonant in the cavity (sample) can be produced. A photodetector placed at a proper position generates a bipolar error signal. The mismatch can be e.g. given by a non-mode-matched input beam [24, 25] or by a tilted input beam [26]. While these methods are easy to implement and low-cost, they are inherently prone to misalignment and to changes of the input beam and the excited cavity transverse mode.

One of the most robust and widely used stabilization schemes today is the *Pound-Drever-Hall locking* scheme, see e.g. [27–29] and references therein. The method was initially developed by Pound [30] for microwave frequency stabilization of and later adapted for optical oscillators [27]. By modulating the input light with an RF source (local oscillator), sidebands of the frequency to be locked are produced. If the distance of the sidebands from the resonant frequency in the frequency domain is large enough so that these sidebands are not resonant, the sidebands can be used as the reference. The signal reflected by the input coupler of the resonant cavity carries the desired phase information and can be used as the sample signal in a phase-sensitive heterodyne detection scheme which demodulates the reflection signal against the RF source [27].

The second stabilization task consists in locking $p_{laser,2}$ to $p_{laser,1}$. As discussed in the beginning of this subsection, this stabilization concerns the laser only. While the repetition frequency of the laser is easy to access as an RF signal from a photo-diode, the detection of the CE phase slippage $\Delta\varphi$ is somewhat more intricate. The most common and straight-forward scheme for the detection of ω_{CE} is based on an f-to-$2f$ interferometer, see e.g. [5] and references therein. Here, an octave-spanning frequency comb is employed to generate a beat signal between the mode with the number N and the one with the number $2N$. Such a broad spectrum can be generated e.g. by self-phase modulation in a nonlinear crystal. The two frequencies $\omega_N = N\omega_r + \omega_{CE}$ and $\omega_{2N} = 2N\omega_r + \omega_{CE}$ generate a beat note with an RF frequency corresponding to their frequency difference, i.e. $\omega_{2N} - \omega_N = \omega_{CE}$. Controlling ω_{CE} can be done in manifold ways. A slow control can be obtained by varying the length of the oscillator beam path through a dispersive material, which influences the ratio of the group to the phase velocity, see e.g. [31]. Recently, a related method was demonstrated in which a composite plate is used to vary this ratio without affecting the repetition rate [32]. Faster controls can be achieved by modulating the power of the pumping beam, see e.g. [19, 20, 33] or by shifting the pumping beam with respect to the cavity mode [34].

Locking a Single Degree of Freedom

The necessity to actively control a second degree of freedom (i.e. the lock of $p_{laser,2}$ to $p_{laser,1}$) mainly depends on the jitter, on the optical bandwidth, on the cavity finesse and on the required stability of the enhanced frequency comb. For many practical applications, fluctuations of two orthogonal degrees of freedom, e.g. of ω_r and ω_{CE} around the state of optimum overlap have a very similar effect on the enhancement, making a second active loop unnecessary. This will be shown quantitatively in the following calculation, in close analogy to Sect. 2.1.2 of our paper [35].

For our derivation, we assume that the central optical frequency ω_c is being locked to the peak of a corresponding cavity resonance by actively controlling solely the repetition frequency ω_r of the comb, and we calculate the effect of a variation $\Delta\omega_{CE}$

(a)

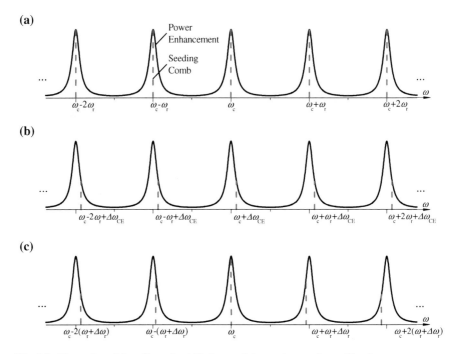

Fig. 2.8 Illustration of the effect of a shift $\Delta\omega_{CE}$ of the carrier-envelope offset frequency ω_{CE} on the enhanced frequency comb, if the laser-cavity lock is implemented by controlling the repetition frequency ω_r so that the central optical frequency ω_c is locked to the peak of a cavity resonance. **a** Optimum overlap. **b** Shift of ω_{CE} by $\Delta\omega_{CE}$. **c** Compensation of $\Delta\omega_{CE}$ through $\Delta\omega_r$ at ω_c

of the other comb parameter ω_{CE} on the enhancement. In the context of the laser-cavity lock the following description of $\theta(\omega)$ is convenient:

$$\theta(\omega) = \tau\omega + \psi(\omega) = 2\pi\frac{\omega}{\omega_r} + \psi(\omega), \qquad (2.26)$$

where $\tau = 2\pi/\omega_r$ is the round trip group delay and $\psi(\omega)$ is a term describing the residual intra-cavity dispersion. For simplicity, here we assume that $\psi(\omega) = 0$ and that an initial ω_{CE} is given, which leads to an optimum overlap of the seeding comb modes with the cavity resonances, see Fig. 2.8a. Let N_c denote the mode number of ω_c, so that

$$\omega_c = N_c\omega_r + \omega_{CE} \qquad (2.27)$$

holds. A variation of the comb parameter ω_r, meaning a change of the comb mode spacing, can be entirely compensated for by the active control. In contrast, a variation $\Delta\omega_{CE}$ of the carrier-envelope offset frequency shifts the entire frequency comb, as depicted in Fig. 2.8b. The active control compensates the drift of ω_c by introducing a variation $\Delta\omega_r$ to ω_r:

$$\omega_c = N_c(\omega_r - \Delta\omega_r) + \omega_{CE} + \Delta\omega_{CE}. \tag{2.28}$$

The last two equations imply:

$$\Delta\omega_r = \Delta\omega_{CE}/N_c. \tag{2.29}$$

While the mode with the number N_c retains its initial frequency, the mode with the number $N_c + m$ will experience a frequency shift equal to $m\Delta\omega_r$. This situation is depicted in Fig. 2.8c.

In Fig. 2.9, the effect of a variation $\Delta\omega_{CE}$, corrected by a variation of ω_r as described above, on the enhanced pulse is plotted for several enhancement factors and several optical bandwidths. As a result of this correction the circulating power decreases and the intra-cavity spectrum is narrowed relative to the case of optimum overlap of the seeding comb with the cavity resonances. The relative changes are plotted as functions of $\Delta\omega_{CE}$ and of $\Delta\omega_{CE}/(2\pi)\Delta\omega_{cav}$, where $\Delta\omega_{cav}$ is the full-width-half-maximum of the cavity resonance. The optical bandwidths used in the calculated examples represent typical values used in femtosecond enhancement cavity experiments. The majority of the experiments presented in this thesis are performed with parameters close to the 7-nm case shown in Fig. 2.9c. Throughout these experiments, the seeding comb could be locked to the cavity by using a single feedback loop, cf. Sect. 4.1. However, a high-finesse enhancement of pulses with broader bandwidths might require a second feedback loop, cf. Sect. 5.1.2.

2.1.4 Transverse Modes, Cavity Scan and Quasi-Imaging

The transverse modes of a resonator (and also of free space or a waveguide) are self-consistent transverse electromagnetic field distributions. A textbook treatment of optical resonator transverse modes is given in Sects. 1.6 and 14.3 of Siegman's book *Lasers* [11]. The most relevant theoretical aspects for the work presented in this thesis are presented in Sect. 2 of the paper [36], see also Chap. 6. In this section we emphasize two further aspects.

Cavity Scan

The scan of the oscillator length generating the seeding pulses around the value of the enhancement cavity length is explained in Sect. 2.3 of the paper [36] (see also Chap. 6) and in Figs. 3 and 5b therein, in the context of adjusting the relative phase between higher-order transverse modes. This is particularly useful to enable the simultaneous resonance of a subset of cavity eigen-modes, which can then be coupled by means of obstacles in the beam path to tailor new low-loss transverse modes. This technique, which we call *quasi-imaging*, is addressed in the next paragraph. Besides enabling quasi-imaging adjustment the scan pattern is also a powerful diagnostic

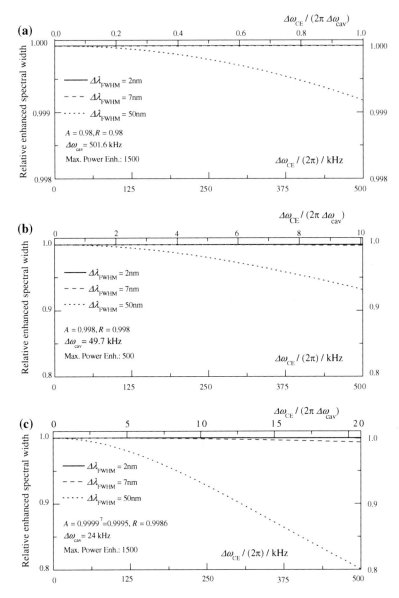

Fig. 2.9 Calculated reduction of the enhanced spectral width (FWHM) after a carrier-envelope offset frequency drift $\Delta\omega_{CE}$ has been compensated for by a shift of the repetition frequency ω_r, locked at the central optical frequency ω_c. The reduction is related to the case of optimum overlap between the frequency comb modes and the cavity resonances. A Gaussian input spectrum is assumed. In a good approximation, the total power enhancement is proportional to the spectral narrowing and the pulse lengthening is inversely proportional to the spectral narrowing. **a** and **b** impedance matched cases with power enhancements at ω_c of 50 and 500, respectively. **c** power enhancement of 1,500 at ω_c, not impedance matched, corresponding to our experimental setup

tool, as discussed in the following.[4] Section 4.2.2 includes examples for all points listed here.

- *Optimum overlap of the axial comb modes of the seeding pulses with cavity resonances.* The scan pattern can be used to adjust the two laser parameters and the cavity parameter manually and coarsely so as to be in the vicinity of the optimum overlap so that subsequent active stabilization ensures a steady state in the optimum overlap, as discussed in the previous section. For sufficiently small cavity dispersion, this is achieved when the main resonance in the scan pattern reaches a maximum and the peaks of the two resonances, one free spectral range to the left and to the right of the main resonance, have equal heights.

- *Transverse mode matching.* The scan pattern also provides information on the amount of light coupled to transverse modes other than the desired one. For instance, in Sect. 3 of [36] the peaks corresponding to higher-order transverse modes are used to calculate the overlap of the input beam with the excited modes. If the target alignment is a perfect transverse mode matching then these peaks must vanish.

- *Error signal adjustment.* As discussed in the previous section, the bipolar error signal for locking should indicate unambiguously the direction and magnitude of the drift which needs to be compensated. Important parameters of the error signal, like amplitude and balancing can be adjusted while scanning over a cavity resonance.

- *Finesse and nonlinearities.* For a given seeding frequency comb, the ratio of the main peak to the ones located a free spectral range to the left and to the right provides a measure for the spectral filtering of the cavity (as discussed in the previous section), which increases with a higher finesse and with dispersion, both of which might also be intensity-dependent. This diagnostic is particularly useful in power scaling experiments, where the (average or peak) power of the seeding comb is increased while all other parameters are kept constant.

- *Stability range detuning and roundtrip phase difference between the sagittal and tangential directions.* As explained in [36], the scan pattern contains information about the detuning from the center of the stability range (see also next paragraph). Moreover, a difference of the Gouy parameters for the x and the y directions in the cavity leads to different on-axis phases[5] of the higher-order modes with the same sum of x and y indices. This results in a multitude of peaks in the scan pattern corresponding to a group of resonances with identical index sum. Conversely, the roundtrip phase difference between the x and the y directions can be determined from this scan pattern (if the finesse is large enough so that these peaks are well resolved). For example, Fig. 5b in Sect. 3 of [36] shows a case in which the Gouy

---

[4] The frequency-domain scan of a Fabry-Perot resonator is not only a helpful diagnostic for the enhancement in this resonator, but a measurement tool in its own right: a cavity with variable length can be used as a passive tunable filter or as a scanning interferometer, see e.g. Sect. 11.5 in [11], scanning a CW laser through a cavity resonance can be employed for ringdown measurements [37] etc.

[5] The accumulation of different on-axis phases manifests itself in different *optical cavity lengths*.

parameters are $\psi_x = 0.78 \cdot 2\pi$ and $\psi_y = 0.76 \cdot 2\pi$. In a stable resonator, in which the beam lies in a single plane, this difference is caused by the difference of the effective radius of curvature of the (de)focusing mirrors thus, indicating the angles of incidence on these mirrors. In an out-of-plane stable resonator this diagnostic can e.g. be used to adjust the angles of incidence on the curved mirrors so that the Gouy parameters in both directions are equal. In this case all peaks corresponding to modes with the same sum of x and y indices coincide. In particular, this could be used to adjust quasi-imaging in both directions, see Sect. 5.2.

Mode Degeneracy and Quasi-Imaging

A phenomenon which can be very useful if controlled properly, and harmful if unwanted, is transverse mode degeneracy. Several transverse modes of different orders can be simultaneously resonant. Intracavity phase front distortions can then lead to a coupling of these modes [38], influencing the excited mode in the cavity.

On the one hand, if fundamental-mode operation is desired this phenomenon leads to a degradation of the beam quality, as discussed in [38]. Such phase front distortions can for instance be induced by thermal or nonlinear effects in the resonator optics. We have observed this in experiments involving an intracavity Brewster plate for XUV output coupling, cf. Sect. 4.4.1. It has also been reported that the gas target can lead to such phase front distortions [39]. The consequence of this unwanted effect is an increase of the roundtrip losses for the distorted mode, possibly a decrease of the overlap with the exciting mode and thus, overall a decrease of the enhancement and of the HHG efficiency. The scan pattern can be used to adjust the relative phases of the transverse modes to avoid resonant coupling among them.[6]

On the other hand, the phenomenon of transverse mode degeneracy, together with the proper choice of obstacles in the beam path and of the seeding transverse mode, enables the controlled combination of eigen-modes of a cavity resulting in a tailored mode with desired properties. An example thereof is the quasi-imaging (QI) concept presented in the paper [36], see also Chap. 6 (and also Sect. 5.2). The initial motivation of the development of QI was to obtain a field distribution which simultaneously exhibits an on-axis maximum for HHG in a gas target close to a cavity focus and, after a certain propagation distance, an on-axis low-intensity region for coupling out the intracavity generated XUV radiation through an opening in a cavity mirror in this region. However, this application can be regarded as a special case of the more general concept of exciting a combination of degenerate transverse modes in a cavity, among which the coupling can be tuned with high precision by means of the position in the stability range. This enables the confinement of different processes to spatially separated regions within the cavity and, at the same time, a precise control of the energy coupling between these regions and therefore between the respective

---

[6] However, the average power during the scan is different from the steady-state regime. Therefore, distortions due to thermal effects might not be visible in the scan pattern.

processes. This concept might prove to be useful for tasks other than providing a direct on-axis access to a high-finesse cavity.[7]

2.2 High-Harmonic Generation in a Gas

For longer than the past two decades the process of high-harmonic generation (HHG) has been extensively studied, both experimentally and theoretically. Owing to its immediate relation to the generation of XUV radiation, such studies have been playing a central role at our institute. In particular, in the context of enhancement cavities, theoretical descriptions of HHG have recently found their way into several PhD theses, e.g. [40–43]. Rather than reproducing in detail the different models describing HHG, this section briefly outlines the semiclassical HHG model known as the *three step model*, and subsequently addresses phase matching considerations, aimed at providing an intuitive introduction to HHG and at enabling the interpretation of the experimental results presented here as well as of the challenges faced towards further power scaling. At the end of the section, we sum up scaling laws for HHG.

2.2.1 The Three-Step Model and Phase-Matching

The Three-Step Model of Single-Atom HHG

The specific structure of high-harmonic spectra generated by multi-cycle laser pulses consists of odd harmonics, whose envelope exhibits: (i) an exponential fall-off (perturbative region) for the lower-order harmonics, (ii) a plateau region of nearly equal power per harmonic, succeeded by (iii) a cutoff in the short-wavelength part of the spectrum. Experimentally, this structure has first been observed in the late eighties [44, 45]. The three step model, developed by Corkum et al., see [46] and references therein, provides a semiclassical, intuitive approach to HHG, explaining many of its properties. In 1994, Lewenstein et al. justified this model with quantum mechanical considerations [47]. In the following, we outline the three step model, explaining the response of a single atom to a high-intensity field. The model is illustrated in Fig. 2.10.

In the first step, the highly intense driving electric field bends the Coulomb potential barrier of the atom so that an electron tunnels from the atomic ground state. A necessary condition for this model to hold is that the so-called Keldysh parameter

---

[7] For instance, it might enable the implementation of the seeding oscillator resonator (with its active medium and mode-locking mechanism) and the passive enhancement resonator in the same cavity (here, *cavity* designates a hollow space rather than a resonator). This could simplify the experimental setup considerably and in particular, spare the necessity of active synchronization.

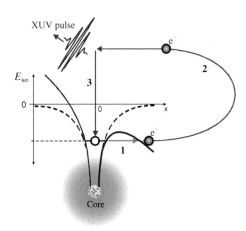

Fig. 2.10 Three step model: *1* tunnel ionization, *2* acceleration through the laser field, *3* recombination to the ground state and emission of an XUV photon. This image is reproduced from [48] by courtesy of Matthias Kling

Table 2.1 Ionization potentials (from Sect. 12.2.1 in [49]).

Gas	Xe	Kr	Ar	Ne	He
U_i/ev	12.13	13.99	15.76	21.6	24.6

$$\gamma = \sqrt{\frac{U_i}{2U_p}} \tag{2.30}$$

is smaller than one [40]. Here, U_i is the ionization potential of the atom, see Table 2.1, and U_p is the ponderomotive energy, i.e. the average kinetic energy of an electron in a harmonic field, given by:

$$U_p = \frac{e^2 E_0^2}{4\omega^2 m_e} \tag{2.31}$$

$$= \frac{e^2}{4\omega^2 m_e} \cdot \frac{2I}{c\epsilon_0}, \tag{2.32}$$

where e, m_e, ω, ϵ_0, E_0 and I are the electron charge and mass, the optical frequency, the vacuum permittivity, the electric field amplitude and the laser beam intensity, respectively. Plugging in the natural constants and using the common measurement units W/cm^2 and μm for I and λ, respectively, yields:

$$U_p \approx 9.33 \cdot 10^{-14} I\lambda^2. \tag{2.33}$$

For $\gamma < 1$ the tunneling times of the electron through the bent Coulomb barrier are short compared to the optical cycle, and the tunnel ionization model implies the formation of a sequence of wave packets, one close to each electric field maximum of the driving multi-cycle laser pulse.

The second step concerns the acceleration of the electron in the laser electric field. It uses classical mechanics to describe well-defined trajectories for the electrons. The trajectories along which the electron wave packets return to the parent ion lead to recombination, which constitutes the third step. Upon recombination, a photon is emitted with an energy equal to the sum of the kinetic energy acquired during acceleration and the ionization potential, i.e. $\hbar\omega = E_{kin} + U_i$. The trajectory and thus, the energy of the emitted photon, depends on the time, i.e. the phase of the driving field, at which the ionization takes place. The model implies that the highest kinetic energy is reached if the electron tunnels at $\omega t \approx 0.3 + k\pi$ rad for any integer k. In this case, the absolute maximum of the kinetic energy amounts to $3.17U_p$, which defines the cutoff frequency via:

$$\hbar\omega_{cutoff} = 3.17U_p + U_i. \tag{2.34}$$

The typical odd-harmonic spectrum originates from the interference of the spectra of the radiation emitted by the different optical cycles of the driving pulse. The cycle-to-cycle repetition of the wave packets passing the parent ions implies that any light that is emitted will be at an odd harmonic of the laser frequency. In the case of single-cycle driving pulses, this interference vanishes and the harmonic spectrum exhibits a rather continuous shape.

Another important result is that only close-to-linear polarization leads the tunneled electron back to the atom and thus to recombination. This is also the key to all isolated as-pulse generation schemes involving polarization gating, see Sect. 5.3.

Multi-Atom HHG: Phase Matching

The response of a macroscopic target, such as a gas jet or cell, to a driving field depends on the spatial phase relation between the radiation contributions of each atom involved in the HHG process, i.e. on phase matching. On the one hand, the coherent addition of these contributions is essential for efficient XUV generation. On the other hand, multiple factors, like the intensity-dependent phases of the excited dipoles, the macroscopic properties of the emitting medium as well as the driving beam space and time properties influence phase matching, making it a complex problem. A thorough treatment of HHG phase matching exceeds the scope of this thesis. Here, the discussion is restricted to an overview of the main factors determining phase matching.

Firstly, the atomic phase of the participating dipoles is intensity-dependent and thus, a function of space and time. Section 11.2 of [49] gives an introduction hereof. An important result is that two geometries are found for which good phase matching is achieved. One of them corresponds to collinear phase matching and results in a Gaussian-like emitted harmonic beam. The other one corresponds to noncollinear phase matching and yields an annular beam.

Secondly, for a Gaussian beam with the confocal parameter

$$b = \frac{2\pi w_0^2}{\lambda}, \tag{2.35}$$

where w_0 is the $1/e^2$-intensity decay focus radius, the geometrical phase shift with respect to a plane wave (Gouy phase shift), depends on both the axial and radial coordinates, z and r, respectively. For a Gaussian beam with a focus at $z = 0$ and a wavefront curvature $R(z)$, the geometrical phase is given by the expression:

$$\phi_{\text{geom}}(r, z) = -\arctan\left(\frac{2z}{b}\right) + \frac{\pi r^2}{\lambda R(z)}. \tag{2.36}$$

The intensity-dependence of the atomic dipole moment together with the geometrical phase lead to a substantial dependence of the coherence properties of the generated harmonics on the position of the nonlinear medium with respect to the focus and on its geometry (see e.g. [50] for details). Assuming a homogenous transverse distribution of the nonlinear medium, the geometry of the interaction region is mainly described by its length l_{med}. If $l_{\text{med}} < b$ holds, the system is referred to as exhibiting *weak (or loose) focusing*. For $l_{\text{med}} > b$ the focusing is said to be *tight*. As a rule, optimum phase matching is reached employing loose focusing conditions [51–53]. The high quality of the transverse mode of an enhancement cavity favors phase matching.

Thirdly, the atomic density plays an essential role in phase matching since dispersion introduces a phase mismatch $\Delta k = k_q - q k_1$ between the phase k_q of the harmonic field of order q and the phase k_1 of the fundamental (polarization) field. In a dispersive medium, the phase lag between the nonlinear polarization introduced by the fundamental laser, oscillating with the frequency ω, at the frequency $q\omega$ and the free Gaussian beam propagating in the medium is [54]:

$$\Delta\phi(z) = (1 - q)\arctan\left(\frac{2z}{b}\right) + \Delta k z. \tag{2.37}$$

The values of Δk for a fundamental wavelength close to the one employed in our experiment are given in Table 1 in [55] for q between 3 and 33. Another important contribution to the phase mismatch is caused by the free electrons of the ionized gas, see also Table 1 in [55]. The influence of several gas target parameters on the harmonic yield is discussed in Sect. 12.2.2 in [49]. In general, an optimum interplay of the mechanisms contributing to phase matching can be found empirically, by adjusting the geometry, the position and the pressure of the gas nozzle.

2.2.2 Scaling Laws

For the design (dimensioning) of any HHG device, the knowledge of the scaling laws of this highly nonlinear process is imperative. However, as opposed to the cases of many other nonlinear processes, in the case of HHG in a gas, a simple and general scaling model, consisting of closed expressions of all relevant experimental parameters, which predict the spectrally resolved intensity of the generated high harmonics, still lacks.[8] This is partly due to the complexity of the process and the multitude of involved parameters, but also due to the absence of HHG experiments enabling the validation of theoretical models over large parameter variation ranges. In particular, the strong intensity-dependence of the process makes the comparison between different experiments difficult. In this section we sum up the most relevant HHG scaling laws, which have been confirmed by all experiments so far.

For a better overview we first list the formulas linking the pulse energy E_{pulse}, the average power P_{av}, the repetition frequency (rate) f_{rep}, the pulse peak power P_{peak}, the pulse duration τ_{pulse} and the peak intensity I of a pulse with a Gaussian distribution in time and space:

$$E_{pulse} = \frac{P_{av}}{f_{rep}}, \tag{2.38}$$

$$P_{peak} = 0.94\frac{E_{pulse}}{\tau_{pulse}}, \tag{2.39}$$

$$I = 2\frac{P_{peak}}{A_{focus}} = 2\frac{P_{av}}{f_{rep}\tau_{pulse}\pi w_0^2}. \tag{2.40}$$

Equations (2.31) and (2.34) imply for the cutoff frequency:

$$\hbar\omega_{cutoff} = 2.96 \cdot 10^{-13} I\lambda^2 + U_i. \tag{2.41}$$

Thus, higher harmonics are generated using a longer driving wavelength. In particular, among the most common mode-locked lasers, this constitutes an advantage of Yb-based systems (centered around 1,030 nm) over Ti:Sa systems (centered around 800 nm). Scaling the cutoff by means of the intensity can be done by (see Eq. (2.40)): (i) increasing the pulse peak power, meaning increasing P_{av} and/or decreasing τ_{pulse} and/or decreasing f_{rep}. The challenges related to these scaling measures from the point of view of enhancement cavities are discussed in Sect. 3.2.1. (ii) by tightening the focusing. However, this usually alters phase matching (see previous section). Moreover, increasing the intensity is limited by ionization. The probability for the latter increases with the intensity, leading to an eventual breakdown of HHG. For example, for Xe the optimum intensity for HHG lies in the range $\sim 10^{13} - 10^{14}$ W/cm^2.

---

[8] First closed-form relations for the conversion efficiency under the assumption of a square-shaped pulse are provided in [56]. However, the focusing conditions and intensity-dependent phase matching are left out of this model.

Furthermore, the cutoff can be pushed towards higher frequencies by choosing lighter atoms with a larger ionization potential (see Table 2.1) since ω_{cutoff} depends on U_i via Eq. (2.34).

Concerning the conversion efficiency, a compendium of several HHG experiments is provided in Sect. 3.1.2 of [57]. A theoretical study [58] for driving lasers from the near-visible (800 nm) to the mid-infrared (2 μm), has recently predicted a dramatic conversion efficiency (CE) improvement for HHG driven by shorter wavelengths:

$$CE \propto \lambda^{-6} \ldots \lambda^{-5}. \tag{2.42}$$

This scaling law has been experimentally verified in the range 800–1850 nm for Xe ($CE \propto \lambda^{-6.3 \pm 1.1}$) and Kr ($CE \propto \lambda^{-6.5 \pm 1.1}$) [59].

In weak focusing operation (i.e. for $l_{\mathrm{med}} < b$), at constant driving intensity, the intensity of plateau harmonics I_{plateau} is observed to scale according to the following law (see Sect. 12.2.1 in [49, 60]):

$$I_{\mathrm{plateau}} \propto b^3. \tag{2.43}$$

The transverse mode characteristics of the generated harmonics depend strongly on the fundamental beam shape, the intensity and the phase matching conditions. In practice, spatial coherence measurements can be performed (cf. e.g. Sect. 12.1.2 in [49]). For HHG with a focused Gaussian beam the following approximation of the divergence of the harmonics holds (cf. Sect. 4.2.1 in [40]):

$$\frac{\theta_{\mathrm{perturb}}^q}{\theta_{\mathrm{f}}} \propto \frac{1}{\sqrt{q}}, \tag{2.44}$$

$$\frac{\theta_{\mathrm{plateau}}^q}{\theta_{\mathrm{f}}} \propto \frac{1}{q}, \tag{2.45}$$

where q, θ_{f}, $\theta_{\mathrm{perturb}}^q$ and $\theta_{\mathrm{plateau}}^q$ denote the harmonic order, the divergence of the fundamental beam with $\theta_{\mathrm{f}} = \lambda/(\pi w_0)$, the divergence of the qth harmonic in the perturbative regime (exponential decay, low harmonics) and the divergence of the qth plateau harmonic, respectively.

Like in the case of the $\lambda^{-(6 \ldots 5)}$ and the b^3-laws in Eqs. (2.42) and (2.43), respectively, formulating a scaling law for the harmonic yield with respect to the pulse duration τ_{pulse} requires a constant intensity. However, unlike in the aforementioned two cases, where the variables λ and b have a well-defined influence on the HHG process, the variation of τ_{pulse} and the subsequent adjustment of the intensity to a constant level, is not a well-defined operation. For instance, τ_{pulse} can be varied by modifying the amplitude spectrum of the pulse and/or the spectral phase. The latter influences the distances between the zeros of the electric field rather than their number, which determines the number of HHG events. Moreover, the *intensity envelope* of the pulse, which can exhibit a complex shape, influences the ionization of the gas *during* the interaction with the pulse, implying saturation and dispersion processes. Thus, the terms *pulse duration* and *intensity* are too coarse to fully describe the multitude of possibilities related to these quantities. Despite this shortcoming several special cases can be found in the literature. In [49, 60] it is stated that the harmonic yield is proportional to τ_{pulse}. This however, follows from the assumption hat the number of XUV photons generated at a given intensity is proportional to the pulse duration, which only holds in the absence of ionization.[9] In [61] and [62] the pulse duration is reduced from 750 to 120 fs and from 200 to 25 fs, respectively, by removing an imposed chirp. In both cases the intensity is kept constant by decreasing the pulse energy along with its temporal compression. While the first publication reports no dependence of the harmonic distribution on the pulse duration, the second one states that shorter pulses lead to an increased conversion efficiency. Moreover, fewer oscillations per pulse, which in the transfer-limited case means shorter pulses, imply broader spectra of the individual harmonics.

References

1. T. Brabec, F. Krausz, Intense few-cycle laser fields: frontiers of nonlinear optics. Rev. Mod. Phys. **72**, 545 (2000)
2. F. Krausz, M. Ivanov, Attosecond physics. Rev. Mod. Phys. **81**, 163 (2009)
3. T. Udem, J. Reichert, R. Holzwarth, T.W. Hänsch, Absolute optical frequency measurement of the cesium D_1 line with a mode-locked laser. Phys. Rev. Lett. **82**, 3568 (1998)
4. T. W. Hänsch, R. Holzwarth, J. Reichert, T. Udem, Measuring the frequency of light with a femtosecond laser frequency comb, in Proceedings of the International School of Physics Enrico Fermi, Course CXLVI, p. 747 (2001)
5. T. Udem, R. Holzwarth, T.W. Hänsch, Optical frequency metrology. Nature **16**, 233 (2002)
6. S.T. Cundiff, Phase stabilization of ultrashort optical pulses. J. Phys. D: Appl. Phys. **35**, R43 (2002)
7. S.T. Cundiff, Femtosecond comb technology. J. Korean Phys. Soc. **48**, 1181 (2006)
8. T.R. Schibli, I. Hartl, D.C. Yost, M.J. Martin, A. Marcinkevicius, M.E. Fermann, J. Ye, Optical frequency comb with submillihertz linewidth and more than 10 W average power. Nat. Phot. **2**, 355 (2008)
9. J.-C. Diels, W. Rudolph, Ultrashort Laser Pulse Phenomena (Academic Press, New York, 1996)
10. C. Rullière, Femtosecond Laser Pulses, Principles and Experiments (Springer, Heidelberg, 2005)
11. A. Siegman, *Lasers* (University Science Books, Sausalito, 1986)
12. H.A. Haus, J.G. Fujimoto, E.P. Ippen, Analytic theory of additive pulse and Kerr lens mode locking. IEEE J. Quantum Electron. **28**, 2086 (1992)
13. E.P. Ippen, Principles of passive mode locking. Appl. Phys. B: Lasers Opt. **58**, 159 (1994)
14. A. Baltuska, M. Uiberacker, E. Goulielmakis, R. Kienberger, V. Yakovlev, T. Udem, T. Hänsch, F. Krausz, Phase-controlled amplification of few-cycle laser pulses. IEEE J. Sel. Top. Quantum Electron. **9**, 972 (2003)
15. L. Xu, C. Spielmann, A. Poppe, T. Brabec, F. Krausz, T.W. Hänsch, Route to phase control of ultrashort light pulses. Opt. Lett. **21**, 2008 (1996)
16. T. Brabec, *Strong Field Laser Physics* (Springer, Cambridge, 2008)

---

[9] For shorter pulses, the saturation intensity is higher, allowing to drive HHG at higher intensities. In a certain sense, this contradicts the τ_{pulse}-proportionality of the harmonics yield.

17. I. Pupeza, X. Gu, E. Fill, T. Eidam, J. Limpert, A. Tünnermann, F. Krausz, T. Udem, Highly sensitive dispersion measurement of a high-power passive optical resonator using spatial-spectral interferometry. Opt. Express **18**, 26184 (2010)
18. J.C. Petersen, A.N. Luiten, Short pulses in optical resonators. Opt. Express **11**, 2975 (2003)
19. K.W. Holman, R.J. Jones, A. Marian, S.T. Cundiff, J. Ye, Intensity-related dynamics of femtosecond frequency combs. Opt. Lett. **28**, 851 (2003)
20. D.R. Walker, T. Udem, C. Gohle, B. Stein, T.W. Hänsch, Frequency dependence of the fixed point in a fluctuating frequency comb. Appl. Phys. B **89**, 535 (2007)
21. R.L. Barger, M.S. Sorem, J. Hall, Frequency stabilization of a cw dye laser. Appl. Phys. Lett. **22**, 573 (1973)
22. T. Hänsch, B. Couillaud, Laser frequency stabilization by polarization spectroscopy of a reflecting reference cavity. Opt. Comm. **35**, 441 (1980)
23. I. Pupeza, T. Eidam, J. Rauschenberger, B. Bernhardt, A. Ozawa, E. Fill, A. Apolonski, T. Udem, J. Limpert, Z.A. Alahmed, A.M. Azzeer, A. Tünnermann, T.W. Hänsch, F. Krausz, Power scaling of a high repetition rate enhancement cavity. Opt. Lett. **12**, 2052 (2010)
24. C.E. Wieman, S.L. Gilbert, Laser-frequency stabilization using mode interference from a reflecting reference interferometer. Opt. Lett. **7**, 480 (1982)
25. M.D. Harvey, A.G. White, Frequency locking by analysis of orthogonal modes. Opt. Commun. **221**, 163 (2003)
26. D.A. Shaddock, M.B. Gray, D.E. McClelland, Frequency locking a laser to an optical cavity by use of spatial mode interference. Opt. Lett. **24**, 1499 (1999)
27. R.W.P. Drever, J.L. Hall, F.V. Kowalski, J. Hough, G.M. Ford, A.J. Munley, H. Ward, Laser phase and frequency stabilization using an optical resonator. Appl. Phys. B: Photophys. Laser Chem. **31**, 97 (1983)
28. E.D. Black, An introduction to Pound-Drever-Hall laser frequency stabilization. Am. J. Phys. **69**, 79 (2001)
29. R. J. Jones, I. Thomann, J. Ye, Precision stabilization of femtosecond lasers to high-finesse optical cavities. Phys. Rev. A **69**, 051803 (2004)
30. R.V. Pound, Electronic frequency stabilization of microwave oscillators. Rev. Sci. Instrum. **17**, 490 (1946)
31. R. Ell, U. Morgener, F.X. Kärtner, J.G. Fujimoto, E.P. Ippen, V. Scheuer, G. Angelov, T. Tschudi, M.J. Lederer, A. Boiko, B. Luther-Davies, Generation of 5-fs pulses and octave-spanning spectra directly from a Ti:sapphire laser. Opt. Lett. **26**, 373 (2001)
32. R. Ell, J.R. Birge, M. Araghchini, F.X. Kärtner, Carrier-envelope phase control by a composite plate. Opt. Express **14**, 5829 (2006)
33. N. Haverkamp, H. Hundertmark, C. Fallnich, H.R. Telle, Frequency stabilization of mode-locked Erbium fiber lasers using pump power control. Appl. Phys. B **78**, 321 (2004)
34. W.-Y. Cheng, T.-H. Wu, S.-W. Huang, S.-Y. Lin, C.-M. Wu, Stabilizing the frequency of femtosecond Ti:sapphire comb laser by a novel scheme. Appl. Phys. B: Lasers Opt. **92**, 13 (2008)
35. I. Pupeza, T. Eidam, J. Kaster, B. Bernhardt, J. Rauschenberger, A. Ozawa, E. Fill, T. Udem, M. F. Kling, J. Limpert, Z. A. Alahmed, A. M. Azzeer, A. Tünnermann, T. W. Hänsch, F. Krausz, Power scaling of femtosecond enhancement cavities and high-power applications, Proc. SPIE **7914**, 791411 (2011)
36. J. Weitenberg, P. Russbüldt, T. Eidam, I. Pupeza, Transverse mode tailoring in a quasi-imaging high-finesse femtosecond enhancement cavity. Opt. Express **19**, 9551 (2011)
37. K. An, C. Yang, R.R. Dasari, M.S. Feld, Cavity ring-down technique and its application to the measurement of ultraslow velocities. Opt. Lett. **20**, 1068 (1995)
38. R. Paschotta, Beam quality deterioration of lasers caused by intracavity beam distortions. Opt. Express **14**, 6069 (2006)
39. T.K. Allison, A. Cingöz, D.C. Yost, J. Ye, Cavity extreme nonlinear optics. in preparation, preprint: arXiv:1105.4195 (2011)
40. C. Gohle, A coherent frequency comb in the extreme ultraviolet. Dissertation, MPQ/LMU, 2006

41. J. Rauschenberger, Phase-stabilized ultrashort laser systems for spectroscopy. Dissertation, MPQ/LMU, 2007
42. A. Ozawa, Frequency combs in the extreme ultraviolet. Dissertation, MPQ/LMU, 2009
43. B. Bernhardt, A. Ozawa, I. Pupeza, A. Vernaleken, Y. Kobayashi, R. Holzwarth, E. Fill, F. Krausz, T.W. Hänsch, T. Udem, Green Enhancement Cavity for Frequency Comb generation in the extreme ultraviolet. CLEO, paper QTuF3, 2011
44. A. McPherson, G. Gibson, H. Jara, U. Johann, T.S. Luk, I.A. McIntyre, K. Boyer, C.K. Rhodes, Studies of multiphoton production of vacuum-ultraviolet radiation in the rare gases. J. Opt. Soc. Am. B **4**, 595 (1987)
45. M. Ferray, A. L'Huillier, X.F. Li, L.A. Lomprk, G. Mainfray, C. Manus, Multiple-harmonic conversion of 1064 nm radiation in rare gases. J. Phys. B: At. Mol. Opt. Phys. **21**, L31–L35 (1988)
46. P.B. Corkum, Plasma perspective on strong-field multiphoton ionization. Phys. Rev. Lett. **71**, 1994 (1993)
47. M. Lewenstein, P. Balcou, M.Y. Ivanov, A. L'Huillier, P.B. Corkum, Theory of high-harmonic generation by low-frequency laser fields. Phys. Rev. A **49**, 2117 (1994)
48. M.F. Kling, M.J.J. Vrakking, Attosecond electron dynamics. Annu. Rev. Phys. Chem. **59**, 463 (2008)
49. P. Jaegle, *Coherent Sources of XUV Radiation* (Springer, New York, 2006)
50. P. Salieres, A. L'Huillier, M. Lewenstein, Theory of high-harmonic generation by low-frequency laser fields. Phys. Rev. Lett. **74**, 3776 (1995)
51. E. Constant, D. Garzella, P. Breger, E. Mevel, C. Dorrer, C.L. Blanc, F. Salin, P. Agostini, Optimizing high harmonic generation in absorbing gases: model and experiment. Phys. Rev. Lett. **82**, 1668 (1999)
52. J.-F. Hergott, M. Kovacev, H. Merdji, C. Hubert, Y. Mairesse, E. Jean, P. Breger, P. Agostini, B. Carre, P. Salieres, Extreme-ultraviolet high-order harmonic pulses in the microjoule range. Phys. Rev. A **66**, 021801(R) (2002)
53. E. Takahashi, Y. Nabekawa, M. Nurhuda, K. Midorikawa, Generation of high-energy high-order harmonics by use of a long interaction medium. J. Opt. Soc. Am. B **20**, 158 (2003)
54. P. Balcou, A. L'Huillier, Phase-matching effects in strong-field harmonic generation. Phys. Rev. A **47**, 1447 (1993)
55. A. L'Huillier, X.F. Li, L.A. Lompré, Propagation eefects in high-order harmonic generation in rare gases. J. Opt. Soc. Am. B **7**, 527 (1990)
56. E.L. Falcao-Filho, V.M. Gkortsas, A.Gordon, F.X. Kärtner, Analytic scaling analysis of high harmonic generation conversion efficiency. Opt. Express **17**, 11217 (2009)
57. J.G. Eden, High-order harmonic generation and other intense optical field-matter interactions: review of recent experimental and theoretical advances. Prog. Quantum Electron. **28**, 197 (2004)
58. J. Tate, T. Auguste, H.G. Muller, P. Salieres, P. Agostini, L.F. DiMauro, Scaling of wave-packet dynamics in an intense midinfrared field. Phys. Rev. Lett. **98**, 013901 (2007)
59. A.D. Shiner, C. Trallero-Herrero, N. Kajumba, H.-C. Bandulet, D. Comtois, F. Legare, M. Giguere, J.-C. Kieffer, P.B. Corkum, D.M. Villeneuve, Wavelength scaling of high harmonic generation efficiency. Phys. Rev. Lett. **103**, 073 902 (2009)
60. P. Balcou, C. Cornaggia, A.S.L. Gomes, L.A. Lompré, A. L'Huillier, Optimizing high-order harmonic generation in strong fields. J. Phys. B: At. Mol. Opt. Phys. **25**, 4467 (1992)
61. K. Kondo, N. Sarukura, K. Sajiki, S. Watanabe, High-order harmonic generation by ultrashort KrF and Ti:sapphire lasers. Phys. Rev. A **47**, R2480 (1993)
62. J. Zhou, J. Peatross, M.M. Murnane, H.C. Kapteyn, Enhanced high-harmonic generation using 25 fs laser pulses. Phys. Rev. Lett. **76**, 752 (1996)

Chapter 3
Objectives of the Experiment and Technological Challenges

The objectives of the experiments presented in this thesis are formulated in Sect. 3.1. The technological challenges on the way to achieving these objectives are reviewed in Sect. 3.2. This section also gives an overview of the solutions developed during the course of this work and of the remaining challenges.

The grand goals of the research in the field of high-repetition-rate, passively-enhanced HHG are pushing the cut-off of the HHG process towards higher energies and scaling up the power and the bandwidth of each harmonic. From the point of view of the enhancement cavity technology, these goals pose analogous challenges as with any other intracavity high-intensity interaction process. Therefore, the objectives formulated in the following can be easily translated to other applications, and the results presented in this thesis have significant implications beyond high-repetition-rate HHG.

3.1 Objectives

A fundamental constraint of the scaling of the enhancement cavity technique towards the goals mentioned above is the ionization of the nonlinear medium, see Sect. 2.2.2. Since the intensity I of the fundamental radiation is given by the ratio of the peak power P_{peak} to the focus size A_{focus}, efficient scaling can only be performed if both are increased. While increasing A_{focus} is a relatively straight-forward cavity-design task, scaling up the peak power is technologically challenging and will be discussed in the following. Moreover, increasing the focus size (i.e. loose focusing) is also desirable due to the b^3-scaling law for the intensity of the plateau harmonics.

Increasing P_{peak} can be achieved by increasing the pulse energy E_{pulse} and/or decreasing the pulse duration τ_{pulse}, which is equivalent to increasing the enhancement bandwidth. For a fixed repetition rate (which is e.g. given by the envisaged application), scaling up E_{pulse} means increasing the average power P_{av}. The

I. Pupeza, *Power Scaling of Enhancement Cavities for Nonlinear Optics*,
Springer Theses, DOI: 10.1007/978-1-4614-4100-7_3,

enhancement-cavity-related challenges concerning the increase of P_{av} and the decrease of τ_{pulse} are discussed in Sect. 3.2.1.

In conclusion, a viable power scaling strategy consists of the following objectives:

- increasing the average power P_{av},
- decreasing the pulse duration τ_{pulse},
- optimizing the HHG process by increasing the focus size A_{focus} according to the scaling of the first two,
- coupling out the intracavity generated harmonics efficiently.

Currently, from the point of view of the seeding lasers, Yb-based systems are the most appropriate sources for pursuing the first two objectives, see also Sect. 5.1.2. Furthermore, Yb-based lasers emit radiation at a longer wavelength than Ti:Sa-systems, which are the closest alternative for seeding high-power femtosecond enhancement cavities. Due to the λ^2-scaling law of the cutoff frequency of the high harmonics, this constitutes another advantage of the Yb-based systems. Conversely, the $\lambda^{-(6...5)}$-scaling law for the conversion efficiency implies a roughly 3 to 4 times lower conversion for Yb-based systems. However, this shortcoming can usually be readily compensated by means of an average power increase. It should also be mentioned that the highly desirable generation of isolated as-pulses offers the most promising prospects in conjunction with short pulses (see Sect. 5.3), which is in accordance with the second objective. Moreover, extending the operation range of cavity-based HHG towards higher average powers and shorter pulses will lead to a better understanding of the HHG process itself by enabling the variation of isolated parameters over larger ranges.

3.2 Challenges

3.2.1 Power and Bandwidth Scaling of an Empty Cavity

Average and Peak Power Scaling for a Constant Finesse

The optics for a given cavity design impose circulating average and peak power limitations. To push the damage threshold of the cavity and enable power scaling these limitations need to be investigated and overcome. One of the major results of this thesis is the investigation of the power scaling of a standard-design bow-tie cavity consisting of commercially available, state-of-the art dielectric mirrors, see Sect. 4.3. In particular, it was found that power scaling is primarily limited by intensity effects in the cavity mirrors. On the one hand, this insight led to the development of next-generation enhancement cavity designs, see Sect. 5.1.1. The design goal of these novel cavities is to increase the spot size on all the mirrors, thus decreasing the intensities on the mirrors. On the other hand, while the novel designs are expected to allow for average and peak power scaling beyond the levels presented in this thesis, they present new challenges. Firstly, average-power-related effects, such as absorption, scattering and thermal lensing are expected to play an increasingly important role. Secondly, increased spot sizes in general request an increased mechanical stability. These aspects are discussed in Sect. 5.1.1. Besides improved cavity designs, mirror development promises to contribute to pushing the current power scaling limits.

Broad Bandwidth Versus High Finesse

Enhancing seeding pulses with shorter durations, i.e. increased bandwidths, represents manifold challenges in a high-finesse cavity. Firstly, non-zero roundtrip group-delay dispersion leads to spectral filtering and chirps magnification in the steady state, see Sect. 2.1.2. Secondly, the broader the bandwidth and the higher the cavity finesse, the larger is the difference between the influence of drifts of the two orthogonal frequency comb degrees of freedom on the comb-cavity overlap, cf. Sect. 2.1.3. Thus, the requirements of the comb-cavity synchronization become more strict.

These effects need to be taken into account when designing the reflectivity and phase properties of the cavity mirrors in the frame of the technological limitations. In particular, trading in power enhancement (or cavity finesse) for a broader circulating bandwidth might be useful. Owing to the relatively narrow optical bandwidth of the pulses used in the experiments presented in this thesis, the effects mentioned above only played a role when intracavity nonlinearities were pronounced. However, for further bandwidth scaling, special attention needs to be paid to these effects, cf. Sect. 5.1.1.

In the course of this thesis a diagnostic method was developed, which allows for the fast and accurate determination of the cavity roundtrip dispersion, including even intracavity nonlinear effects. This method (described in Sect. 4.2.2) is expected to be instrumental in the design process of mirrors for future bandwidth scaling.

3.2.2 Inclusion of a Gas Target

The mode of operation of femtosecond enhancement cavities relies on the super-position of the intracavity circulating pulse with every pulse of the seeding train at each bounce off the input coupler. Any distortion of the intracavity pulse affects this superposition and therefore, the performance of the cavity. From this point of view, nonlinear cavity dynamics [1–4] and in particular, the interaction of the fundamental radiation with the gas target, require special attention.

Very recently, two publications have tackled the still open question of the effect of the intracavity plasma on the cavity performance [3, 4]. The majority of the experimental findings reported there have been qualitatively observed in our system as well (see e.g. Sect. 3.3 in the paper [5]). In our group a thorough analysis of these effects remains a subject for future study. In the following we review the most important laser-gas interaction mechanisms influencing the performance of the cavity and constituting future challenges related to power scaling.

Effect on Longitudinal Modes

The interaction of the high-intensity fundamental radiation with the gas affects both the amplitude and the phase of the circulating field. The amplitude effect is due to the energy which the laser loses when producing photoelectrons and increases if the interaction volume is increased, see e.g. [6].

There are two processes with characteristic timescales on the order of the pulse repetition period [3]: (i) the plasma decay and (ii) the travel time of gas atoms in the interaction region. If the pulse repetition period is shorter than these two times, the number of ionized atoms and of electrons in the interaction region will be large and dispersion will affect the circulating beam, cf. Sect. 2.2.1. Moreover, ionization is accompanied by self-phase modulation (SPM) and a spectral blue-shift, resulting in additional dispersion and affecting the spectral overlap at the input coupler.

To avoid ionization and enable efficient HHG from a large interaction volume, finding the optimum intensity will be necessary and will presumably play a more important role than in experiments with smaller interaction volumes. Moreover, the prevention of cumulative effects becomes more important as the interaction volume increases, due to the longer travel time of gas atoms through the high-intensity region.

Effect on Transverse Mode

The plasma also acts as a negative lens, defocusing the beam and influencing the excited cavity transverse mode [3]. If several transverse modes are degenerate (or close to degeneracy), this phase front distortion might lead to a coupling among them, as explained in Sect. 2.1.4. To avoid this, a proper position in the stability range, where coupling among modes is strongly suppressed and/or spatial filters might be useful. Moreover, this effect might be detrimental to the quasi-imaging technique (see Sect. 2.1.4), where the coupling among modes of different transverse orders is crucial.[1]

Effect on Lock Stability

As discussed above, the nonlinear cavity response due to the gas target has a dynamic nature. This might lead to locking instabilities and/or irreproducibility, a fact which has been observed experimentally in our system and in [3, 4]. One of the main reasons for this is that the error signal for the lock of the cavity parameter and one of the laser parameters (see Sect. 2.1.3) is usually generated from an isolated (and narrow)

---

[1] However, in principle it could be pre-compensated for by proper transverse mode matching.

part of the enhanced spectrum. In principle, the narrower this optical bandwidth, the more accurate is the confinement of the error signal generation to a spectral region, which ideally extends over a single cavity resonance. However, an extremely narrow bandwidth also means a poor SNR. Therefore, in practice this bandwidth extends over many cavity free spectral ranges, which, in the case of a nonlinear response, fluctuate according to the dynamic nature of the optical dispersion.

Usually, the the error signal for locking is generated using the fundamental beam itself. If the effect of the plasma on the transverse mode exhibits dynamic behavior, this might lead to error signal distortions. A solution to this problem might be the use of an intermediary single-frequency continuous-wave (CW) laser, which is coupled into the cavity in addition to the seeding frequency comb [7]. Locking both the cavity and one seeding laser parameter to this CW reference promises several advantages. First, the SNR of the error signal is dramatically increased due to the confinement of the laser power generating the error signal to a single frequency [8]. Second, SPM will not affect the CW beam. And third, the fluctuations of the spatial profile of the resonant CW beam will carry information on the time scale(s) of plasma dynamics.[2]

3.2.3 XUV Output Coupling

Finding a suitable output coupling solution has been one of the main tasks within the HHG enhancement cavity community over the past few years and also a subject of this thesis. The strict design criteria of an XUV output coupler makes this task highly challenging. This challenge is described in the first part of Sect. 2 of the paper [9] (on page 12110), see also Chap. 6. The results regarding output couplers obtained during the work that led to this thesis are discussed in Sect. 4.4. An outlook on the development of XUV output coupling techniques planned in our group is provided in Sect. 5.2. An overview of the methods proposed and demonstrated so far is given in Appendix 7.3.

References

1. K.D. Moll, R.J. Jones, J. Ye, Nonlinear dynamics inside femtosecond enhancement cavities. Opt. Express **13**, 1672 (2005)
2. V.L. Kalashnikov, Femtosecond pulse enhancement in an external resonator: impact of dispersive and nonlinear effects. Appl. Phys. B **92**, 19 (2008)
3. T.K. Allison, A. Cingöz, D.C. Yost, J. Ye, Cavity extreme nonlinear optics. in preparation, preprint: arXiv:1105.4195 (2011)
4. D. Carlson, J. Lee, J. Mongelli, E. Wright, R. Jones, Intracavity ionization and pulse formation in femtosecond enhancement cavities. Opt. Lett. (2011)

---

[2] Probing the negative plasma lens to investigate whether the plasma decays between the pulses or not, could in principle be also done by sending a single-pass beam trough the focus and observing its phase front distortions.

5. I. Pupeza, X. Gu, E. Fill, T. Eidam, J. Limpert, A. Tünnermann, F. Krausz, T. Udem, Highly sensitive dispersion measurement of a high-power passive optical resonator using spatial-spectral interferometry. Opt. Express **18**, 26184 (2010)
6. M. Geissler, G. Tempea, A. Scrinzi, M. Schnürer, F. Krausz, T. Brabec, Light propagation in field-ionizing media: extreme nonlinear optics. Phys. Rev. Lett. **83**, 2930 (1992)
7. R. Holzwarth, personal communication
8. S. Holzberger, personal communication
9. I. Pupeza, E. Fill, F. Krausz, Low-loss VIS/IR-XUV beam splitter for high-power applications. Opt. Express **19**, 12 108 (2011)

Chapter 4
Experimental Setup and Results

Lehrling ist Jedermann. Geselle, der was kann. Meister, der was ersann.

Alter Handwerkerspruch

4.1 Yb-Based CPA System with 80 MHz, 200 fs, 50 W

The laser system providing the pulses for our enhancement cavity was designed and built by the fiber laser group from the Institute of Applied Physics at the University of Jena. The main components, the mode of operation and the performance are described in detail in [1]. This section concentrates on the adaptations made to the laser system, which were necessary for the operation in conjunction with our enhancement cavity.

4.1.1 The Oscillator

The 78 MHz-repetition-rate pulse train is delivered by an Yb:KYW solid-state bulk crystal based, passively mode-locked oscillator, depicted in Fig. 4.1. The crystal is optically pumped by a Jenoptik CW laser Diode providing 2.2 W of average power at a central wavelength of 980 nm for a pump current of 3.5 A. The central wavelength of the transfer-limited 170 fs pulses is 1042 nm. The average output power of the oscillator for a diode pump current of 3.5 A is 150 mW. This value deviates from the 220 mW reported in [1] due to the changes made to the oscillator, which are addressed in the following.

As discussed in Sect. 2.1.3, for resonant enhancement the two independent frequency comb parameters need to be set in the vicinity of the optimum overlap with the enhancement cavity resonances. Then, at least one of these parameters needs to

I. Pupeza, *Power Scaling of Enhancement Cavities for Nonlinear Optics*,
Springer Theses, DOI: 10.1007/978-1-4614-4100-7_4,
© Springer Science+Business Media New York 2012

Fig. 4.1 Oscillator picture (see text for details)

be actively locked using a feedback loop. We manually vary the oscillator cavity dispersion by displacing a fused silica wedge in the beam. A second wedge is used to account for the direction change of the beam passing through the first wedge, see Fig. 4.1. The angle of incidence on the wedge surface is close to Brewster's angle to ensure low losses for p polarization. Varying the roundtrip dispersion of the oscillator cavity changes the difference between the phase and the group velocity and thus, influences ω_{CE} and the carrier-envelope offset slippage, cf. Sect. 2.1.3 an in particular Eq. (2.4). In addition, we can vary the current driving the pump diode to influence the laser parameters. An active control of this degree of freedom is not installed yet. However, it could be implemented in future for a fast stabilization of the second comb parameter see e.g. [2–4]. The position of the wedge as well as the pump current are both set manually. Once the seeding frequency comb is brought in the vicinity of the optimum overlap with the cavity resonances by means of at least one of these two mechanisms, we lock the laser to the cavity with a fast piezo-electric transducer (PZT), onto which one of the cavity end mirrors is glued. We found empirically that the amplitudes of the drifts of the two laser parameters are small enough so that stable locking over several minutes can be performed with a single parameter, as described in Sect. 2.1.3.

The mirror used for the active lock is a semiconductor saturable absorbing mirror (SESAM, provided by Amplitude Systems), which is also responsible for the

Fig. 4.2 SESAM-PZT-recoil
body construction

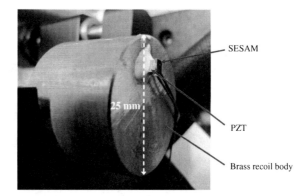

SESAM

PZT

Brass recoil body

mode-locking. The SESAM surface was initially $\sim5\times5\,\mathrm{mm}^2$. We chose this mirror to implement the active stabilization because it is the thinnest (and therefore lightest) one in the oscillator cavity and also because the impinging beam diameter is the smallest of all mirrors. The SESAM was cut into 4 equal square pieces, one of which was glued on a fast PZT (PI-PL022.31), see Fig. 4.2. The opposed surface of the PZT was glued onto a bulky brass recoil body to avoid mechanical resonances of the system at acoustic frequencies. The maximum travel of the PZT is $2.2\,\mu\mathrm{m}$. To ensure that the PZT system does not have mechanical resonances at low acoustic frequencies, we applied a sinusoidal signal to the PZT and slightly misaligned the oscillator cavity, so that a PZT travel resulted in an amplitude change of the oscillator output signal. We swept the frequency of the driving sinusoidal signal from a few Hz to many kHz and observed the phase shift of the oscillator output modulation with respect to this signal on an oscilloscope. Figure 4.3 shows oscilloscope screenshots for 100 Hz and 29 kHz. While at low frequencies the oscillator output amplitude modulation follows the PZT driving signal with a constant phase, at roughly 30 kHz it exhibits a phase shift of approximately π indicating the presence of a mechanical resonance. The phase shift between the two signals gives a good hint at the presence of mechanical resonances so that this method can be applied to roughly determine the linear range of the mechanical system's transfer function. In this case we can assume a linear response up to approx. 20 kHz. While our locking loop performs satisfactorily for enhancement factors of just below 2,000, the system is very sensitive to acoustic noise and vibrations in this high-finesse operation regime. An improvement of the lock stability is one of the near-term goals in the further development of this system. To increase the available lock bandwidth, new mirror-on-PZT systems could be used, such as the recently demonstrated system with a linear bandwidth in excess of 180 kHz, see [5].

The oscillator was empirically found to be the most critical component in terms of mechanical stability. Therefore, additional measures were taken to avoid both fast and long-term drifts, cf. Fig. 4.1. To suppress sound waves inside the oscillator box, we attached absorption panels (Axifoam SH001, 50 mm) to the inner side of the surrounding box walls. Moreover, to avoid a feedback of the pump diode light, a

Fig. 4.3 Bandwidth measurement of the PZT construction shown in Fig. 4.2, performed by a misalignment of the oscillator so that **a** the output beam is modulated according to the displacement of the PZT. *Red* PZT driving voltage, *blue* oscillator output. **b** A phase shift of $\sim\pi$ is observed around 30 kHz, indicating the first mechanical resonance

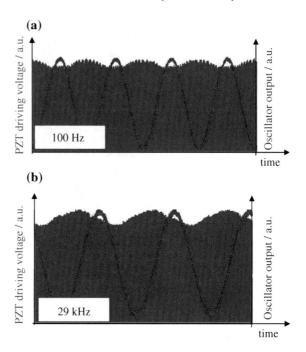

dichroic mirror was used in the resonator. The transmitted pump light was dumped onto a bulky beam dump for thermal stability.

4.1.2 Chirped-Pulse Amplification

The pulses generated by the oscillator are stretched to 150 ps in a transmission-grating based stretcher (1,250 lines/mm) with an efficiency of 55 %. Subsequently, the chirped pulses are amplified in a two-stage fiber amplifier using polarization-maintaining photonic-crystal fibers, see [1]. The amplified pulses are compressed with two highly efficient fused silica transmission gratings with 1,250 lines/mm [6]. The output pulse parameters are 200 fs and up to 60 W of average power. The deviation with respect to the parameters published in [1] is due to the better alignment of the oscillator, resulting in a broader bandwidth and due to several exchanged components, including the fibers and the pump diodes.

By varying the distance of the two compressor gratings, the output pulse duration can be adjusted continuously between the close-to-transfer-limited value of 200 fs and several ps while leaving the other output parameters constant. The dependence of the spectral phase on the grating distance can be found e.g. in Appendix A.2 of [7]. The straightforward variation of the output pulse duration over a large range is particularly important for peak power scaling experiments at constant average powers.

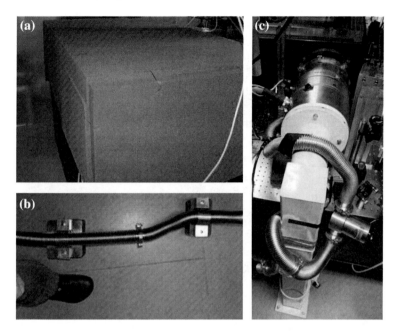

Fig. 4.4 **a** Protection housing of the laser against mechanical vibrations and air fluctuations. **b** Vibration-damping heavy iron blocks attached to the bellows connecting the pre-vacuum pump with the turbo pump. **c** Mounting of the turbo pump and connection to the vacuum chamber containing the enhancement cavity (see text for details)

4.2 The Enhancement Cavity

4.2.1 Setup

The enhancement cavity is described thoroughly in the papers [8–10], see also Chap. 6. In this section the description is rounded up with additional details on the passive protection of the setup against mechanical vibrations and heating. Measures against these factors include the use of rigid components (base plates, mounts etc.) i.e. with large inertia, the use of oscillation damping mounting posts and a housing around the laser system, doubled with absorption panels (Axifoam SH001, 50 mm), see Fig. 4.4a. The latter also prevents air fluctuations. Scattered and residual light is dumped using cooled beam dumps to avoid heating of the system components which might lead to instabilities. Despite these measures, a long-term drift of the relative length between the seeding oscillator and the enhancement cavity can be observed. Its origin is subject to current investigation.

The vacuum pumps attached to the chamber containing the cavity constitute an unavoidable source of vibrations and we took several measures to damp these vibrations before they reached the vacuum chamber. The pump building up the pre-vacuum

(Busch Fossa 0030A/B scroll pump) is attached to a turbomolecular drag pump (Pfeiffer TMH 1000M) with $\sim$4 m long flexible bellows. Several heavy iron blocks are attached to the bellows to damp the vibrations caused by the pre-pump, see Fig. 4.4b. A vibration damping bellows is mounted between the turbo pump and the vacuum chamber. The pump itself is fixed to an iron pillar attached to the floor, see Fig. 4.4c. The vacuum chamber walls are also mechanically detached from the base plate of the enhancement cavity so that the coupling of vibrations from the vacuum pumps is minimized.

4.2.2 Diagnostics

The diagnostics used to characterize the intracavity beam are addressed in the papers [8–11], see also Chap. 6. This section contains additional details.

The intracavity beam is mainly monitored by observing the leakage through the highly reflecting cavity mirrors. The devices used to this end are: a power meter (Coherent OP-2 IR, FieldMate), two optical spectrum analyzers (Ocean Optics HR 4000 and ANDO AQ 6315A), an autocorrelator (APE PulseCheck), a photodiode (Hamamatsu InGaAs G8376) and a CCD camera (WinCam D). The transmission of the cavity mirrors used for diagnostics amounts to 1.65×10^{-6} when measured with an incident power of 50 W (estimated uncertainty of 10 %). The transmission and reflectivity are nearly constant over the entire bandwidth of the enhanced pulses. However, when many kW of power circulate inside the cavity, this transmission value might change due to effects caused by the large powers. Thus, in order to prevent systematic errors in the intracavity beam characterization, we implemented a second measurement method of a different nature to the first one which uses the leakage beam. In the cavity around the laser beam, we placed a circular aperture with a diameter around 3 times larger than the $1/e^2$ beam diameter (see Fig. 1.b on page 2052 of the paper [8]). This aperture is realized as a hole in a reflector. It clips the beam and sends the clipped portion to a detector. In theory, the beam intensity profile after one cavity round-trip behind this aperture (i.e. the far field of the aperture) corresponds to the convolution of the Fourier transform (FT) of the beam profile just before the aperture with the FT of the aperture function (Fraunhofer diffraction). With the aperture being very large in respect to the beam size, the FT of its function can be approximated by a very narrow peak, so that the beam profile remains largely unaffected by this convolution. In practice, we choose the diameter empirically as small as possible under the constraint that the enhancement is not significantly affected by the aperture. Even though the clipped portion of the beam is insignificant for the cavity enhancement, owing to the large circulating powers, it is well measurable. Since the ratio of the powers measured with both methods (i.e. leaking portion and clipped portion) stayed constant throughout the power scaling experiments [8], we assume that all measurements using the leakage beam are reliable.

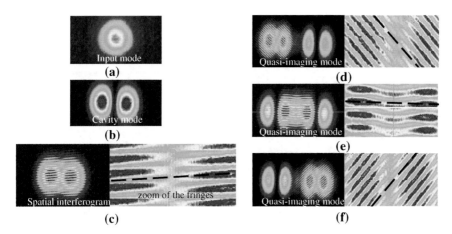

Fig. 4.5 Spatial interference of a copy of the input beam **a** with higher-order modes, detected on a CCD camera, **b** the cavity TEM_{10} mode, **c** spatial interferogram of the beams in (**a**) and (**b**). The *dashed line* indicates that the lobes of the excited mode are anti-phased, **d–f** interferograms of a quasi-imaging mode. The two central lobes are in phase (**e**) while the first two (**d**) and the last two (**f**) lobes are anti-phased

A diagnostic method, which we developed in the course of this work is the cavity roundtrip dispersion measurement via spatial-spectral interferometry of a copy of the seeding laser beam with a copy of the intracavity circulating beam. The method is described in the paper [9], see also Chap. 6. By exchanging the imaging spectrometer with a CCD camera, the interferometer is also suited for the detection of the spacial phase relation between the lobes of higher-order modes excited in the cavity. Examples are shown in Fig. 4.5. For best fringe contrast, the delay of the two interferometer arms was set to 0.

One of the most useful diagnostics for the alignment of the system is the periodic scan of the seeding comb through the cavity resonances, see also Sect. 2.1.4. Ideally, the scan time over a resonance should be chosen much longer than the cavity build-up and ring-down times, so that at each point in the scan a close-to-stationary state can be assumed. However, unless the relative positions of the comb and the cavity are actively locked during this scan, in a real experiment fluctuations are likely to distort the scan patt rn if the scan speed is too slow. This holds in particular for high-finesse cavities with long build-up times. Therefore, for the diagnostic methods described in the following, we choose a scan speed which is large enough to prevent fluctuations rom prevailing and slow enough to allow for a clear distinction of peaks with different magnitudes in the scan pattern. In our system, usually, the scan is performed with a triangle signal with 20 Hz period and a PZT excursion of up to 2.2 μm.

As mentioned in Sect. 2.1.4, the intracavity intensity evolution while scanning the frequency comb over the cavity resonances (i.e. the scan pattern) is a powerful diagnostic. The scan pattern can be used to monitor changes of the cavity finesse, the

(a)   **(b)**

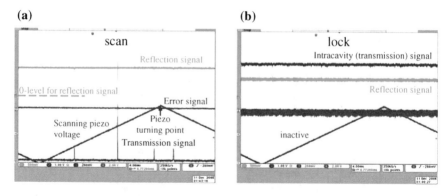

Fig. 4.6 Oscilloscope screenshots of: **a** cavity length scan (amplitude slightly larger than $2FSR$, period 20 Hz). The *central* peak corresponds to the optimum overlap of the comb modes with the cavity resonances. The peaks to the *left* and to the *right* correspond to the cases in which the comb is shifted by approx. one FSR with respect to the optimum overlap by means of the repetition rate (secondary resonances). Due to good transverse mode matching, no light is coupled to higher-order transverse modes. **b** The locked state. Here, an effective power enhancement of 700 was achieved

enhanced bandwidth, the detuning from the stability range center and the coupling to higher-order modes, all of which is particularly useful in power scaling experiments.

Figure 4.6a shows an oscilloscope screenshot of a scan over slightly more than two FSR. The PZT in the oscillator cavity is driven periodically with a voltage which is proportional to the purple signal. The blue signal indicates the intracavity power, measured with a photodiode through one of the cavity mirrors. The central (highest) peak corresponds to the main resonance, i.e. the case of optimum overlap of the seeding comb with the cavity resonances. The secondary peaks to the left and to the right are generated by a comb-cavity overlap which is reached when the change of the repetition rate is large enough so that it has a similar effect to an overall shift of the frequency comb of approximately one cavity FSR. However, these peaks are smaller and broader than the main resonance peak due to the repetition rate mismatch. Increasing the repetition rate (i.e. cavity length) mismatch further leads to further secondary resonances. The corresponding peaks exhibit decreasing amplitudes and increasing widths as the mismatch grows. The ratio of the maxima of the main resonance peak to the secondary resonances increases with the coupled bandwidth. In the case of a CW laser, all resonances have the same hight.

Figure 4.7 shows two scan patterns which were recorded with a 1-mm thick fused silica intracavity Brewster plate. Except for the input pulse duration, all seeding beam parameters are kept the same in the measurements in panels (a) and (b). The fact that the central peak (main resonance) is significantly weaker for input pulses with an autocorrelation duration of (a) $\tau_{AC} = 0.3$ ps than for (b) $\tau_{AC} = 2.4$ ps clearly indicates nonlinear effects limiting the intracavity peak power.

The scan pattern is also an indicator for the transverse mode matching and the alignment. For example, the alignment which led to Fig. 4.6a is very good, since higher-order modes are barely visible. In contrast, intensity peaks corresponding to

(a)   **(b)**

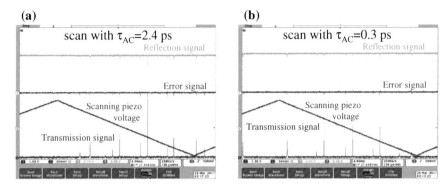

Fig. 4.7 Oscilloscope screenshots of the scan patterns with a 1-mm thick fused silica intracavity Brewster plate, for an input power of 35 W and two different pulse autocorrelation durations: **a** $\tau_{AC} = 0.3\,ps$ and **b** $\tau_{AC} = 2.4\,ps$. The peaks corresponding to higher-order modes are clearly visible, which indicates a suboptimal alignment. The changing peak maximum ratio of the *central* peak to the next secondary peaks confirms the nonlinear effects observed in the other measurements (see text)

higher-order-modes are visible in the scan patterns shown in Fig. 4.7, indicating a misalignment either of the input beam or of the cavity (see also Sect. 4.4.1).

Moreover, the scan pattern can be used to coarsely adjust the two laser parameters and the cavity parameter in a vicinity of the optimum overlap, so that the active feedback loop(s) can subsequently provide a locked steady state. Firstly, it is necessary that the main peak lies in the scan range of the PZT. This can be adjusted by manually setting the relative length of the oscillator and the enhancement cavity. Secondly, the other laser parameter can be coarsely adjusted (e.g. by means of the wedges in the oscillator cavity or the pump power of the oscillator). In our case, the empty cavity dispersion over the enhanced optical bandwidth is small enough so that it does not affect the circulating pulse significantly. In this case, optimum overlap is achieved when the central peak is maximized and, in a good approximation, the closest secondary resonances have equally high peaks. For example, this is the case in Fig. 4.6a but not in Fig. 4.7a.

Another important scan pattern signal monitors the reflection of the input beam from the cavity input coupler. As the frequency comb approaches optimum overlap during the scan, light is coupled to the cavity, which results in an intensity dip in the reflection signal, cf. Figs. 4.6a, 4.7 and 4.8a. The dip depth and shape primarily depend on the transverse mode matching, the spatial region of the beam profile which is used to generate the signal, the scan speed and the impedance matching. Let us first assume perfect transverse mode matching, see e.g. Fig. 4.8a. The scan speed determines the time the light has to build up in the cavity and hence, influences the hight of the transmission signal peak(s) and the depth of the reflection dip(s), thereby giving a measure of the amount of light coupled into the cavity. The dependence of the dip depth on the scan speed relates the build-up process to the scan time and could in principle be used to determine the degree of impedance matching of the

(a)   **(b)**

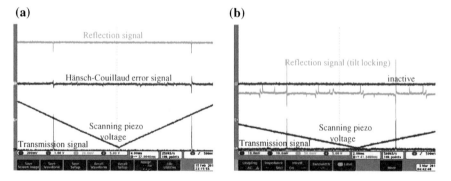

Fig. 4.8 Oscilloscope screenshots of: **a** cavity length scan. The seeding transverse mode is matched to the excited cavity mode so that the reflection signal has a dip shape. In contrast, the Hänsch–Couillaud error signal exhibits a slope when passing through the cavity resonance, whose sign indicates the scan direction (bipolarity). **b** By introducing a transverse mode mismatch of the seeding beam with respect to the cavity mode, the reflection signal (offset for clarity) can be used to produce a bipolar error signal (see text)

cavity. However, we have not done this so far. On the other hand, we frequently use the reflection signal to maximize the dip depth for the coarse setting of the laser and cavity parameters in order to achieve the desired vicinity of the optimum overlap. It is noteworthy that like the transmission signal, the reflection signal in the case of perfect transverse mode matching does not provide a bipolar signal which could be used as an error signal for active stabilization, see e.g. Fig. 4.8a. However, if a slight transverse mismatch of the input beam with respect to the cavity mode is introduced, like a tilt or beam parameter mismatch, then a portion of the reflected beam having a bipolar shape on resonance can be spatially filtered for detection. The reason for this shape is the interference of the reflected part of the input beam that is resonant with the cavity mode (and therefore bears the information on the cavity roundtrip phase) with a portion of the input beam which is spatially filtered out, i.e. rejected by the cavity, and travels along a slightly different path which introduces a delay towards the first part and acts as a reference. This leads to a transverse mode mismatch locking error signal as can be seen in Fig. 4.8b.

Although the above mentioned method was successfully employed to lock the cavity, for most of the experiments presented in this thesis we used the Hänsch–Couillaud locking scheme. Figure 4.8a shows a typical Hänsch–Couillaud signal. With such a signal and an alignment leading to a scan pattern similar to Fig. 4.6a, a stable lock over several minutes can be achieved, see Fig. 4.6b. The scan pattern is also useful for balancing the difference photodiode in the Hänsch–Couillaud scheme as well as the amplitude of the error signal. The transmission signal level difference of the main peak in Fig. 4.6a and the locked level in Fig. 4.6b is due to the scan speed, which at this finesse does not allow for a full build-up of the main resonance. For locking we used a customized proportional integral controller (Menlo Systems Lockbox PIC 201).

Further diagnostics customized during this work include a z-scan setup and a ring-down loss-meter, adapted for arbitrary angles of incidence. These devices are described in the paper [11], see also Chap. 6. The vacuum z-scan setup can be used for nonlinear refractive index measurements (cf. [12, 13]) and as a single-pass diagnostic for nonlinearities in cavity optics. It was developed with Jan Kaster, who also constructed it.

4.3 Power Scaling of the Empty Cavity

One of the major results of this work consists in the power scaling experiments performed with the standard-design bow tie cavity presented in Sect. 4.2.1 and is documented in the paper [8], see also Chap. 6. The significance of these experiments is twofold. On the one hand, they show that the enhancement cavity technique enables the combination of high pulse energies (up to the the mJ level) with high repetition rates (80 MHz) for sub-ps pulses with standard optics. This power regime is ideally suited for driving low-conversion-efficiency nonlinear processes and cannot be achieved with alternative techniques to this day. On the other hand, these experiments reveal power scaling limitations of standard bow-tie-design enhancement cavities built with state-of-the-art mirrors. Identifying these limitations allows for the development of new solutions to overcome them. In particular, the high-intensity induced nonlinear effects on the mirrors have been identified as one of the main power scaling limitations. As a consequence, our group is developing a novel cavity design supporting increased spot sizes (thus decreasing the intensity) on the mirrors, see Sect. 5.1.1. Peak power scaling limitations are identified also in the experiments reported in Sect. 3.2 of the paper [9], see also Chap. 6. By using the novel spatial-spectral-interferometry-based technique, an intensity-dependence of the empty cavity roundtrip *GDD* was measured.

Another phenomenon observed at large peak powers, which is not accounted for in [8] and [9] is third-harmonic generation (THG) in the cavity mirrors. Figure 4.9a shows a picture of a piece of paper placed behind a cavity mirror. The beam transmitted through the mirror contains a portion of UV light, which leads to fluorescence on the paper. With an UV spectrometer, this light was identified as the third harmonic of the infrared fundamental radiation. The spectrometer counts at 347 nm for several input power values at a constant input pulse duration of 200 fs are plotted in Fig. 4.9b, c. The 2.5-power exponential dependence revealed by the log–log plot in Fig. 4.9c deviates from the expected power of three. This deviation most likely stems from the fact that due to THG the circulating power does not scale linearly with the input power, which provides the abscissa of this measurement. Optical THG at interfaces between materials with different refractive indexes is a well-known phenomenon [14]. Most likely, the UV light is generated at the interface vacuum-mirror surface and/or at the interfaces of the coating layers. Independent measurements with the z-scan setup (see Sect. 4.2.2) showed that a portion of this generated third harmonic is transmitted and a portion reflected by the mirror. Therefore, the

(a)   **(b)**   **(c)**

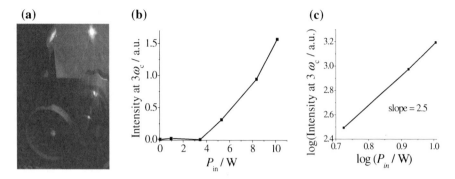

Fig. 4.9 Third-harmonic generation at the cavity mirrors: **a** photograph of cavity mirror surface and of the fluorescence on a piece of paper placed behind the mirror; **b** UV spectrometer counts [a.u.] at 347 nm versus input power P_{in} at a constant input pulse duration of 200 fs; **c** log–log plot of the points, at which THG was detected. The slope indicates a 2.5-power dependence, see text

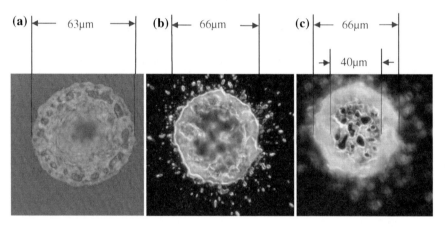

Fig. 4.10 Optical microscope images of enhancement cavity mirror damages: **a** input coupler; **b** and **c** same cavity high reflector; two different focusing depths of the optical microscope

measured THG signal behind the cavity mirrors is a superposition of contributions from several cavity mirrors, each with a slightly different angle of incidence and spot size, which makes tracing back these contributions an involved task.[1] The results plotted in Fig. 4.9b show qualitatively that THG in the cavity mirrors is a potential limitation of this cavity design. This corroborates the need for novel cavity designs with large spot sizes on the mirrors, like the ones described in Sect. 5.1.1.

---

[1] In the frame of this thesis, the THG in the cavity mirrors was not investigated quantitatively. The main reason for this is the characteristic of the cavity mirrors at the third harmonic of the fundamental radiation. Since the employed mirrors are dielectric quarter-wave stacks, they have a strong reflection band in the vicinity of the third harmonic. However, the exact position of this band in the frequency domain depends on the mirror dispersion and also strongly depends on the angle of incidence.

Finally, Fig. 4.10 shows mirror damages produced when driving the cavity with powers resulting in intensities beyond the identified damage thresholds. The study of these damages could perhaps be useful on the way to developing mirrors with higher damage thresholds.

4.4 XUV Output Coupling Techniques

Coupling out the intracavity generated XUV radiation is one of the major challenges in the field of enhancement-cavity-based HHG. In the frame of the work that led to this thesis, several existing output coupling methods were investigated and further developed and also novel techniques emerged. This section addresses this progress. A summarizing overview of the proposed and demonstrated output coupling methods is given in Appendix 7.3.

4.4.1 Brewster Plates

The first method ever employed for coupling out intracavity generated XUV radiation (and still the most widespread method) relies on a thin plate placed under Brewster's angle (for the fundamental radiation) between the HHG focus and the subsequent cavity mirror [15–17]. Since in the XUV the refractive index of the plate material is smaller than one, the harmonic beam experiences total internal reflection at the plate surface and is coupled out of the cavity. In our setup, several thin plates/foils were tested. Figure 4.11a[2] shows the position of the Brewster plate in the cavity and Fig. 4.11b, c show our mount. Note that no measures for cooling the plates are taken. In our group, an extensive material search was conducted by Oleg Pronin with the purpose of identifying the materials suitable as intracavity Brewster plates. The main results of intracavity power scaling experiments with the most promising plates are summarized in Table 4.1.

The effects limiting the power scalability of the cavity including a Brewster plate can be divided into thermal and nonlinear effects. Thermal effects can be mainly attributed to the absorption of the circulating power in the plate. However, in the results presented here, the possibility of scattered light heating up the mount of the plates cannot be excluded. Especially in the case of ultrathin foils spanned in a frame this might quickly lead to damage. To minimize the absorption in the plates, the thickness should be chosen as small as possible. However, this also implies poor heat conduction (see Sect. 5.2.1) and mechanical fragility. The thermal effects mentioned in Table 4.1 manifest themselves in two ways. Firstly, if the position in the stability range of the cavity is set so that several transverse modes are (close to) simultaneously resonant, a coupling between those modes can be observed after

---

[2] The graphic illustration was done by Jan Kaster.

Table 4.1 Power scaling results with different intracavity Brewster plates. P_{circ} denotes the circulating average power

Material	Si$_3$N$_4$	SiC	MgO	C(singlecrystal & CVD)	Al$_2$O$_3$	Sio$_2$ crystaline	Sio$_2$ fused	Sio$_2$ fused
Thickness	30–500 nm	300 nm	30 μm	20–100 μm	340 μm	100–150 μm	100 μm	1000 μm
Maximum $P_{circ}(kW)$	<3	0.5	0.5	–	0.5	5	12	4.5
Enhancement	685	50	500	–	30(?)	400	500	100–250
Observations	Damage at P_{circ} > 3kW	Thermal effects, Strong SHG, no damage	Thermal effects damaged	no resonances observed at all	SHG, thermal effects	Thermal effects, SHG, no damage	Thermal effects, damage after several minutes	Stronger nonlinear effects, no damage

CVD chemical vapor deposition. Thermal and nonlinear effects are discussed in the text

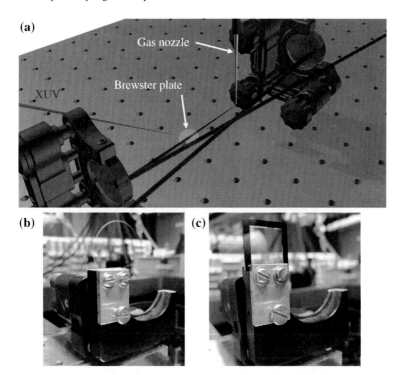

Fig. 4.11 **a** The Brewster plate is located about half way between the cavity focus and the subsequent curved mirror. **b** Brewster plate mount with an 100 μm thin fused silica plate and **c** with an 1 μm thin Si$_3$N$_4$ foil, mounted on a Si frame

a "heating time" on the order of 1 s. In other words, immediately after the cavity is locked, the fundamental, i.e. the GH_{00} mode is excited and shortly after, the transverse field distribution changes, which can be detected with the CCD camera. This effect can be attributed to a phase front distortion in the plate originating either from a thermal deformation of the plate or from a temperature-dependent refractive index. Secondly, thermal effects can be identified via a "cool-down" when switching from the lock to the scan mode after a longer period of locking. Then, the initial position of the main resonance in the scan pattern (cf. Fig. 4.6) just after switching to the scan mode, changes slightly to a steady position, indicating a change of the cavity length. Both phenomena occur on a ∼1 s time scale, leading to the conclusion that they cannot be attributed to nonlinearities. For the thinner plates/foils, thermal effects eventually lead to damage. The nonlinear effects mentioned in Table 4.1 include mainly second harmonic generation (SHG) and self phase modulation (SPM). The latter affects the roundtrip phase of the cavity and leads to spectral filtering which can be identified in the intracavity spectrum and in the overall power enhancement. To distinguish between thermal and nonlinear effects, the input pulses can be chirped while keeping the pulse energy constant.

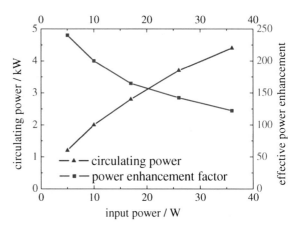

Fig. 4.12 Circulating power and effective power enhancement measured as functions of the input power with a 1-mm thick fused silica intracavity Brewster plate

Fused Silica, 1 mm

As an example of power scaling with an intracavity Brewster plate, we discuss the results with a 1-mm thick fused silica plate. Figure 4.12 shows the measured circulating power and effective power enhancement factor as a function of the input power. The input pulse duration was kept constant (autocorrelation time 300 fs). A substantial power enhancement decrease is observable as the input power increases. To investigate the origin of this effect we recorded the intracavity spectrum for different input powers. Figure 4.13 shows the results. Panel (a) shows the input spectra for 5 W and 35 W at a constant input pulse duration. The power-dependent modulations in the spectrum are due to higher-order modes excited in the main amplifier fiber, which is a sign of suboptimal alignment of the input coupling in this fiber. However, this artefact does not affect the conclusions of this measurement. We observe that even at a relatively moderate power (5 W, purple curve), the dispersion of the Brewster plate is not negligible since only a part of the spectrum is coupled to the cavity. Moreover, this part can be selected by varying the carrier-envelope phase slippage of the input comb (by displacing the wedges in the oscillator cavity), see green curve. As the input power is increased to 35 W, the circulating spectrum becomes narrower, which indicates nonlinear dispersion. To ensure that the observed effects are caused by the peak power, we kept the input power of 35 W constant and measured the circulating spectrum for two different input pulse durations, obtained by chirping the input pulses (but keeping their bandwidth constant). The results are shown in panel (b) of Fig. 4.13. The narrowing of the enhanced spectrum for a measured input pulse autocorrelation of 300 fs as opposed to 2.4 ps is clearly visible and confirms the nonlinear dispersion. The scan pattern shown in Fig. 4.7 provides another confirmation of the nonlinearity (further explanations of this scan pattern can be found in Sect. 4.2.2).

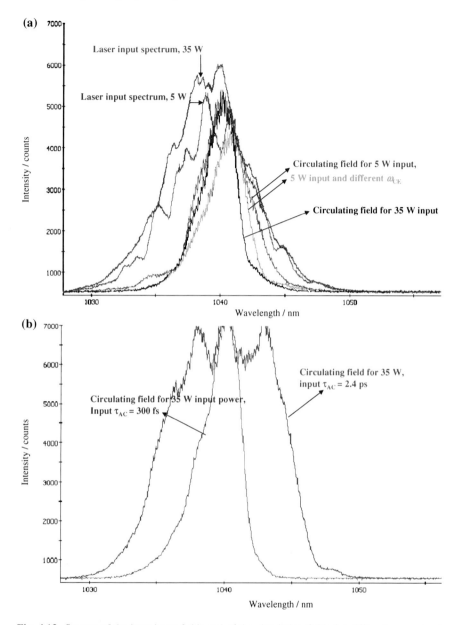

Fig. 4.13 Spectra of the laser input fields and of the circulating fields for different power levels with a 1-mm thick fused silica intracavity Brewster plate (screenshots): **a** spectra for a constant input pulse duration (autocorrelation of 300 fs) and **b** spectra for a constant input power of 35 W and two different input pulse durations. (The high-frequency ripples are caused by the etalon effect of transmission neutral density filters we used to attenuate the beam. They are not related to any cavity-relevant effect.)

Fused Silica, 0.1 mm

The best results with a Brewster plate in our setup were achieved with a $100\,\mu$m thin SiO_2 (fused silica) plate. The introduced dispersion can be neglected for 200 fs circulating pulses. With an enhancement of 500 we were able to stably lock 12 kW of circulating power. However, this regime could only be kept constant for $\sim$5 min, after which time the plate got damaged. This was reproduced several times, using new plates. With a circulating power level of 3 to 4 kW long-time stable operation can be achieved, without damaging the plate. These observations agree well with a similar system reported in [18] and A. Ozawa, personal communication. The results presented in Sect. 4.5.3 were achieved with this plate.

4.4.2 WOMOC: Wedge-On-Mirror Output Coupler

A novel optical element for output coupling, based on the Brewster plate, but exhibiting a better thermal conductivity was developed as part of this thesis and is described in the publication [9], see also Chap. 6. A prototype of the wedge-on-mirror output coupler (WOMOC) was produced and characterized.

A further conceivable manufacturing option for a WOMOC, which is not discussed in [9] is an "upside-down" production, in which the highly reflecting multilayer structure is coated on a bulk substrate, out of which the wedge is subsequently lapped (and polished). For mechanical stability, a supporting substrate can be glued or optically contacted to the last layer of the dielectric coating before processing the upper layer.

Another advantage of the robustness of this element is that an additional (optional) heat conducting structure could be attached to the front surface of the WOMOC by coating, gluing or optical contacting. It should also be mentioned that the mirror underlying the wedge layer need not be a dielectric mirror. In principle, other mirror technologies could be used to realize the same spatial separation effect caused by the wedged layer.

4.4.3 Direct On-Axis Access

Ideally, the intracavity generated XUV radiation does not interact with any material while it is coupled out, a situation to which we refer as *direct* (or *geometrical*) output coupling. Two different direct output coupling techniques are being pursued in our group: (i) the enhancement of the fundamental mode with a small on-axis hole in the mirror succeeding the HHG focus and (ii) quasi-imaging (QI), i.e. the cavity operated with a combination of transverse modes, with a hole (or slit) in the same mirror.

The first approach is discussed in [19]. With the model developed there, a power enhancement of a few hundred is expected with holes with an aperture diameter of

$\sim 100\,\mu$m in our present setup. On the one hand, a precise prediction of the enhancement is difficult because usually the edge of such a hole is imperfect, leading to additional phase front distortions which affect the enhancement. On the other hand, manufacturing such mirrors is highly challenging. The accurate drilling of such a small hole has been demonstrated only recently by Dominik Esser and coworkers at the Fraunhofer ILT in Aachen, by using laser micromachining [20] (see also Sect. 5.2.2).

The major expected disadvantage of the first method with regard to power scaling is the fact that the field distribution of the excited cavity mode has a maximum at the position of the hole. Imperfections of the hole edge might lead to additional absorption and might decrease the damage threshold of the mirror. The quest for an alternative direct output coupling technique, which enables an intensity minimum at an aperture in the mirror following the HHG region (with an on-axis maximum) led to the development of quasi-imaging. This technique is introduced in the publication [10], see also Chap. 6, where a successful proof of principle is also presented. In Sect. 5.2.2 we give an outlook on the steps taken to show that this is a viable XUV output coupling method.

Due to the decreasing of the beam divergence with the harmonic order, cf. Eqs. (2.44, 2.45), the efficiency of direct output coupling increases with the harmonic order. Moreover, since the harmonics do not interact with any material while being coupled out, we consider this to be the most promising strategy for coupling out short-wavelength XUV radiation.

4.4.4 Other Methods

We address two further methods which were developed in close collaboration with our project. However, these methods have not yet been implemented in our experiment.

Reflective Nanograting

A recently developed output coupler relies on a diffractive nanostructure written in the last layer of the dielectric multilayer coating of the cavity mirror following the HHG focus. The nanograting diffracts the XUV light, while the element still acts as a high reflector for the fundamental IR beam. This concept was first presented in [21]. In our group, Ying-Ying Yang and coworkers optimized, produced and characterized such an element. The results are presented in the publication [11], see also Chap. 6.

A main property of the nanograting is the spatial dispersion of the output coupled high harmonics. However, this property is undesirable for several applications, such as isolated attosecond pulse generation (see e.g. Sect. 5.3). Moreover, in the experiments reported in [11] we have found that the nanostructure enhances nonlinear effects, which might limit the power scaling of an enhancement cavity containing such an element.

GIP: Grazing-Incidence Plate

An extension of the Brewster plate method towards larger output coupling efficiencies was developed by Oleg Pronin et al. [22] in our group. By applying an s-polarization broadband anti-reflection coating for large angles of incidence (e.g. 75°) on both sides of a thin transparent plate, this element can be placed in the cavity similarly to the Brewster plate, but under a larger angle of incidence. While the output coupling efficiency is expected to be increased considerably, this approach does not overcome the inherent average (and peak) power scaling limitations of the free-standing Brewster plate. In particular, the coating requires a certain substrate thickness due to mechanical tension.

4.5 XUV Generation

The first XUV measurements made with the system described here were carried out at relatively low powers with a conventional Brewster plate output coupler. There are three reasons for this choice. Firstly, this technique is well understood and allows the XUV detection hardware to be set up. Secondly, measurements made with the Brewster plate can be used as benchmarks for alternative output coupling methods. And thirdly, at the time of writing this thesis, none of the preferred alternative output couplers was available yet (see also Sect. 5.2). In the following we describe the gas target used for HHG and the XUV diagnostics, and show the first XUV spectra recorded with our setup.

4.5.1 Intracavity Gas Target

Figure 4.14 shows the gas nozzles available for our XUV experiments. All the experiments presented in this section were carried out with tapered glass capillaries like the one shown in Fig. 4.14a with Xenon being used as the nonlinear medium. We chose Xe due to its low ionization potential (cf. Table 2.1), enabling HHG at relatively low intensities. Figure 4.14b[3] shows a bright plasma, generated when such a nozzle is placed at a lateral distance of approximately 1 mm from the cavity focus. The brightness of the plasma indicates strong ionization of the gas. On the one hand, this confirms the high intensity in the cavity focus. On the other hand, the ionization dramatically decreases the HHG efficiency. We found out empirically that moving the gas target very close to the beam increases the generated XUV power, see Sect. 4.5.3 and [23]. In this case, the brightness of the plasma decreases significantly. The gas dynamics as well as related phase matching conditions require further intensive investigations, see also Sect. 3.2.2.

---

[3] This picture was taken by Thorsten Näser.

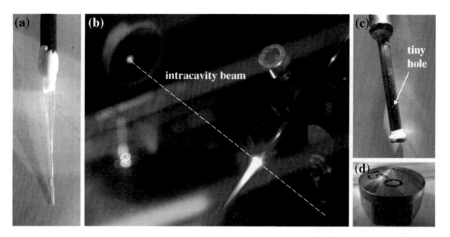

Fig. 4.14 Gas nozzles for HHG: **a** A (hand-made) tapered glass capillary. Typical hole diameters lie between 100 and 200 μm. **b** Bright Xe plasma generated with the glass capillary at a lateral distance of ~1 mm from the focus. **c** Pierced tube (200-μm-diameter holes). **d** 150-μm-diameter Laval nozzle. See text for details

These handmade glass nozzles are cheap, easily produced and have several advantages regarding intracavity HHG. Their alignment is straightforward, requiring no special orientation (at least when the fundamental transverse mode is used for HHG). Moreover, they can be positioned very close to the beam. In contrast, the pierced-tube nozzle shown in Fig. 4.14c is quite likely to clip the beam slightly due to its tiny holes, which are necessary for optimum gas flow. Owing to the large circulating power levels, even absorption of only several ppm due to clipping can lead to overheating and damage of the nozzle material, as was observed in our experiment. The use of a Laval nozzle, as the one shown in Fig. 4.14d will be a subject of future experiments.

Without the gas target, the roughing pump alone provides a vacuum of 0.4 mbar. Switching on the turbo pump yields a pressure of 10^{-5} mbar. The backing pressure of the gas target can be varied up to approximately 4 bar. At a backing pressure of approx. 1.2 bar, which was empirically found to be the optimum for the experiment presented in Sect. 4.5.3, the background pressure in the vacuum chamber was 7×10^{-3} mbar.

4.5.2 XUV Diagnostics

For spectrally resolved measurements we use a McPherson grazing incidence monochromator, model 248/310G. Several XUV diffraction gratings are available, of which we used the one with 133.6 grooves/mm in the experiments presented in this section. The XUV detection can be carried out by scanning a solar blind channel electron multiplier (channeltron) along the Rowland circle of the monochromator.

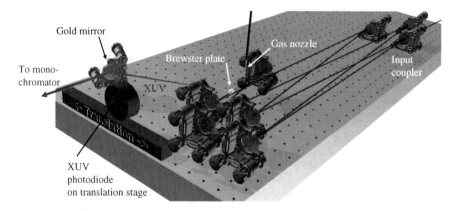

Fig. 4.15 HHG generation and detection setup, see text for details. A close-up of the region around the focus is shown in Fig. 4.11a

We use a McPherson model 425 channeltron. At the time of writing this thesis, detection with a multichannel plate (MCP), model XUV-2040, which is also sensitive to IR radiation, was unsuccessful due to saturation by residual IR. For an absolute XUV power measurement we mounted an XUV photodiode (IRD, model AXUV 100 AL3), shielded against IR radiation with 150 and 300 nm thick Al filters. However, both filters were destroyed by the residual IR radiation, superimposed to the XUV beam. To solve this problem and enable both the MCP and the photodiode measurements, we plan to use additional Brewster plates, or, for a better XUV reflectivity, grazing-incidence optics with an anti-reflection coating for the IR (like the plate presented in [22]) for splitting the residual IR from the output coupled XUV.

Figure 4.15[4] shows a schematic of the XUV generation and detection setup, which was designed to offer maximum flexibility. The XUV beam coupled out by the Brewster plate is steered by a gold mirror to the entrance slit of the monochromator. Figure 4.16a shows the power attenuation encountered by the XUV beam upon reflection at the Brewster plate and at the Au mirror as well as after having propagated through the residual Xe along the distance of 65 cm from the output coupler to the spectrometer. The reflectivity values were calculated for an angle of incidence of 56°, corresponding to Brewster's angle for fused silica with a refractive index of 1.45. The three combined power attenuation contributions are plotted in Fig. 4.16b. To enable a convenient switch between a spectrally resolved measurement and an integrated power measurement, we placed the XUV photodiode on a motorized translation stage, so that it can be readily driven in and out of the beam. This setup can easily be adapted to suit the related output coupling methods of the WOMOC and the GIP.

---

[4] The graphic illustration was done by Jan Kaster.

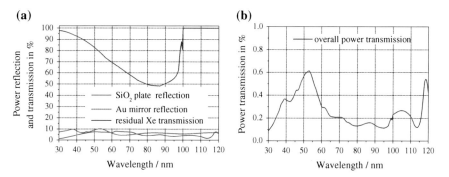

Fig. 4.16 XUV power transmission from the focus to the spectrometer entrance: **a** separate contributions of the fused silica Brewster plate, of the *gold* steering mirror (both calculated with refractive index data from [24], for p polarization) and of the propagation along 65 cm of free space at a surrounding Xe pressure of 7×10^{-3} mbar (data from [25]); **b** combined contributions

4.5.3 XUV Generation: First Results

Locked Cavity

In order to lie below the damage threshold of the free-standing 0.1 mm fused silica Brewster plate, we seeded the cavity with an average power of 7 W. With an input coupler with $R = 99.74\,\%$ and 7 highly reflective mirrors, we measured a circulating average power level of 2.4 kW at the beginning of the experiment. The measured effective power enhancement was 340 with an intracavity pulse autocorrelation identical to the one of the input pulse. We adjusted the cavity close to the center of the stability range, yielding a focus radius $w_0 \approx 23\,\mu\text{m}$. This corresponds to an intensity of $2 \times 10^{13}\,\text{W/cm}^2$ in the focus. The XUV yield was optimized by aligning the nozzle very close to the focus and by adjusting the backing pressure. The optimum for the latter was found to be approx. 1.2 bar. The measurement took several minutes, during which the intracavity power decreased to 1.8 kW, corresponding to a power enhancement of 260 and an intensity of $1.5 \times 10^{13}\,\text{W/cm}^2$ (the measurement was performed starting from long towards short wavelengths). Since the cavity without the Brewster plate can operate steadily for tens of minutes, the most likely cause of the circulating power decrease is the heating of the plate. Further experiments need to be performed in order to gain more insight into this process (see also Sect. 5.2). Figure 4.17a shows the channeltron counts as the wavelength was scanned. Note that these are raw measurement data, i.e. no calibration of the detector was performed. Harmonics up to the 17th order can be clearly observed. The roll-off of the noise level could be attributed to the slow power decrease discussed above. The broad width of the harmonics is caused by the relatively low resolution of the measurement (determined by the monochromator entrance slit size and the employed XUV diffraction grating). The cutoff frequency calculated with Eq. (2.34) and with an average power of 2 kW (which is the value at the respective measurement point) corresponds to a

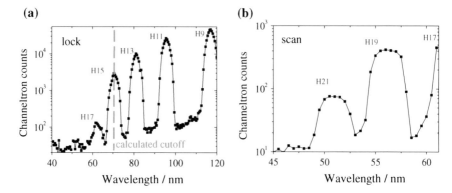

Fig. 4.17 First XUV spectra recorded from the system presented in this thesis. **a** Locked state with 2.4 kW of circulating power (which decreased in time to 1.8 kW), 200 fs intracavity pulse duration. *Green dashed line* calculated cutoff. See text for further details. **b** Spectrum recorded while scanning through the main resonance with 50 W of input power

wavelength of 70 nm, see green dashed line in Fig. 4.17a. The strongly decreased intensity of the 17th harmonic indicates good agreement with the expected cutoff value.

Scan Mode

To detect even higher harmonics without damaging the Brewster plate, we scanned the frequency comb through the matched cavity resonance periodically with an input power of 50 W. In this way high intensities build up, but a large thermal load is avoided since the cavity is off-resonance most of the time. Figure 4.17b shows the results for a 20 Hz scan over approximately half of the cavity *FSR* around the resonance (and with an increased channeltron counts integration time). In this regime, XUV up to the 21st harmonic could be detected. While this mode of operation of the cavity does not serve the purpose of most envisaged experiments, it reveals that the achievable high intracavity power levels bear the potential of generating higher harmonics and more power per harmonic. This enforces the justification to search for better suited output couplers, which was done in the course of this thesis.

Lewenstein Model Simulation

For a further theoretical confirmation of this measurement, the harmonic spectrum was simulated according to the Hydrogen-like atom model developed by Lewenstein et al. see Sect. III.D in [26]. The measured driving laser parameters from the beginning of the measurement (i.e. an intensity of 2×10^{13} W/cm^2) were plugged in the simulation and perfect phase matching was assumed. The latter assumption is

Fig. 4.18 XUV spectrum calculated with the HHG model
developed by Lewenstein et
al. [26], see text for further
details

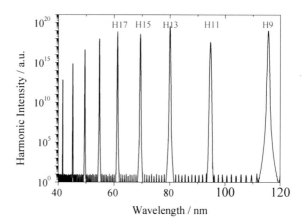

justified by the loose focusing on the one hand and by the empirical optimization of
the gas pressure on the other hand. Figure 4.18 shows the result of the simulation.[5]
The cutoff frequency is at the ∼17th harmonic. Despite this slight mismatch, which
is due to the power decrease during our measurement, this result corroborates the
agreement between HHG theory and our measurement.

References

1. T. Eidam, F. Röser, O. Schmidt, J. Limpert, A. Tünnermann, 57 W, 27 fs pulses from a fiber
   laser system using nonlinear compression. Appl. Phys. B **92**, 9 (2008)
2. K.W. Holman, R.J. Jones, A. Marian, S.T. Cundiff, J. Ye, Intensity-related dynamics of femtosecond frequency combs. Opt. Lett. **28**, 851 (2003)
3. N. Haverkamp, H. Hundertmark, C. Fallnich, H.R. Telle, Frequency stabilization of mode-
   locked Erbium fiber lasers using pump power control. Appl. Phys. B **78**, 321 (2004)
4. D.R. Walker, T. Udem, C. Gohle, B. Stein, T.W. Hänsch, Frequency dependence of the fixed
   point in a fluctuating frequency comb. Appl. Phys. B **89**, 535 (2007)
5. T.C. Briles, D.C. Yost, A. Cingöz, J. Ye, T.R. Schibli, Simple piezoelectric-actuated mirror
   with 180 kHz servo bandwidth. Opt. Express **18**, 9739 (2010)
6. T. Clausnitzer, J. Limpert, K. Zöllner, H. Zellmer, H.J. Fuchs, E.B. Kley, A. Tünnermann,
   M. Jupe, D. Ristau, Highly efficient transmission gratings in fused silica for chirped-pulse
   amplification systems. Appl. Opt. **42**, 6934 (2003)
7. G. Stobrawa, "Aufbau und Anwendungen eines hochauflösenden Impulsformers zur Kontrolle
   ultrakurzer Laserimpulse, Dissertation, Friedrich-Schiller-Universität Jena, 2003
8. I. Pupeza, T. Eidam, J. Rauschenberger, B. Bernhardt, A. Ozawa, E. Fill, A. Apolonski,
   T. Udem, J. Limpert, Z.A. Alahmed, A.M. Azzeer, A. Tünnermann, T.W. Hänsch, F. Krausz,
   Power scaling of a high repetition rate enhancement cavity. Opt. Lett. **12**, 2052 (2010)
9. I. Pupeza, X. Gu, E. Fill, T. Eidam, J. Limpert, A. Tünnermann, F. Krausz, T. Udem, Highly
   sensitive dispersion measurement of a high-power passive optical resonator using spatial-
   spectral Interferometry. Opt. Express **18**, 26184 (2010)

---

[5] This simulation was carried out by Ernst Fill.

10. J. Weitenberg, P. Russbüldt, T. Eidam, I. Pupeza, Transverse mode tailoring in a quasi-imaging high-finesse femtosecond enhancement cavity. Opt. Express **19**, 9551 (2011)
11. Y.-Y. Yang, F. Süssmann, S. Zherebtsov, I. Pupeza, J. Kaster, D. Lehr, E.-B. Kley, E. Fill, X.-M. Duan, Z.-S. Zhao, F. Krausz, S. Stebbings, M.F. Kling, Optimization and characterization of a highly-efficient diffraction nanograting for MHz XUV pulses. Opt. Express **19**, 1955 (2011)
12. M. Sheik-Bahae, A.A. Said, E.W.V. Styrland, High-sensitivity, single-beam n_2 measurements. Opt. Lett. **14**, 955 (1989)
13. M. Sheik-Bahae, A.A. Said, T.H. Wei, D.J. Hagan, E.W.V. Styrland, Sensitive measurement of optical nonlinearities using a single beam. IEEE J. Quantum Electron. **26**, 760 (1990)
14. T.Y.F. Tsang, Optical third-harmonic generation at interfaces. Phys. Rev. A **52**, 4116 (1995)
15. C. Gohle, T. Udem, M. Herrmann, J. Rauschenberger, R. Holzwarth, H.A. Schuessler, F. Krausz, T.W. Hänsch, A frequency comb in the extreme ultraviolet. Nature **436**, 234 (2005)
16. R. Jones, K.D. Moll, M.J. Thorpe, J. Ye, Phase-coherent frequency combs in the vacuum ultraviolet via high-harmonic generation inside a femtosecond enhancement cavity. Phys. Rev. Lett. **94**, 193201 (2005)
17. A. Ozawa, J. Rauschenberger, C. Gohle, M. Herrmann, D.R. Walker, V. Pervak, A. Fernandez, R. Graf, A. Apolonski, R. Holzwarth, F. Krausz, T. Hänsch, T. Udem, High harmonic frequency combs for high resolution spectroscopy. Phys. Rev. Lett. **100**, 253901 (2008)
18. A. Ozawa, Y. Kobayashi, Intracavity high harmonic generation driven by Yb-fiber based MOPA system at 80 MHz repetition rate. CLEO, paper CThB4 (2011)
19. K. Moll, R. Jones, J. Ye, Output coupling methods for cavity-based high-harmonic generation. Opt. Express **14**, 8189 (2006)
20. D. Esser, W. Bröring, J. Weitenberg, H.-D. Hoffmann, Laser-manufactured mirrors for geometrical output coupling of intracavity-generated high harmonics. manuscript in preparation
21. D.C. Yost, T.R. Schibli, J. Ye, Efficient output coupling of intracavity high harmonic generation. Opt. Lett. **33**, 1099–1101 (2008)
22. O. Pronin, V. Pervak, E. Fill, J. Rauschenberger, F. Krausz, A. Apolonski, Ultrabroadband efficient intracavity XUV output coupler. Opt. Express **19**, 10232 (2011)
23. C. Altucci, T. Starczewski, E. Mevel, C.-G. Wahlström, B. Carré, A. L'Huillier, Influence of atomic density in high-order harmonic generation. J. Opt. Soc. Am B **13**, 148 (1996)
24. E.D. Palik, G. Ghosh, E.J. Prucha, *Handbook of Optical Constants of Solids* (Academic Press, New York, 1985)
25. Center for X-ray Optics (http://www.cxro.lbl.gov)
26. M. Lewenstein, P. Balcou, M.Y. Ivanov, A. L'Huillier, P.B. Corkum, Theory of high-harmonic generation by low-frequency laser fields. Phys. Rev. A **49**, 2117 (1994)

Chapter 5
Outlook

5.1 Further Power Scaling of Enhancement Cavities

The power scaling results presented in Sect. 4.3 have shown that the enhancement in a standard-design passive cavity, built with state-of-the art, commercially available ion-beam sputtered dielectric multilayer mirrors, is primarily limited by nonlinear effects in the mirrors. An approach to circumventing this limitation consists in increasing the laser spot size on the mirrors, thus decreasing the incident intensity and relaxing nonlinear effects in the mirrors. Moreover, a larger irradiated area decreases the thermal gradient which is beneficial for the high-average-power regime. In this section we first report on two large-mode-area (LMA) enhancement cavity designs, in close analogy to [1], developed in our group as a consequence of the power scaling results presented in this thesis. Then we address two laser sources which will be employed for the next power scaling experiments. Combining these laser sources with the next-generation LMA cavities offers on the one hand the potential of scaling up the average power of femtosecond laser pulse enhancement to the MW level and on the other hand provides the prospect of few-cycle IR pulse enhancement.

5.1.1 Novel Cavity Designs

We discuss three ways of increasing the laser spot sizes on the mirrors of ring resonators that can be implemented separately or combined. The first method relies on increasing the angle of incidence (AOI) on the mirrors. The laser spot size increases proportionally to the reciprocal cosine of the AOI leading to a decrease of the intensity by the same factor. This approach can be implemented in a ring resonator configuration with oblique incidence (OI) on the mirrors. The radiation can be focused by two consecutively arranged off-axis parabolic mirrors, as shown in

I. Pupeza, *Power Scaling of Enhancement Cavities for Nonlinear Optics*,   73
Springer Theses, DOI: 10.1007/978-1-4614-4100-7_5,
© Springer Science+Business Media New York 2012

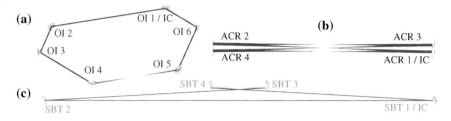

Fig. 5.1 a Oblique incidence setup: OI 1 is the input coupling (IC) mirror, OI 4 and OI 5 are identical high reflecting (HR) off-axis parabolas and OI 2-4 are flat HR mirrors. **b** All-curved-mirror resonator: ACR 1 is the IC, the other mirrors are HR. All mirrors are curved with the same radius. **c** Standard bow tie setup: SBT 1 is a flat IC, SBT 2 is a flat HR mirror and SBT 3 and SBT 4 are curved HR mirrors with the same radius

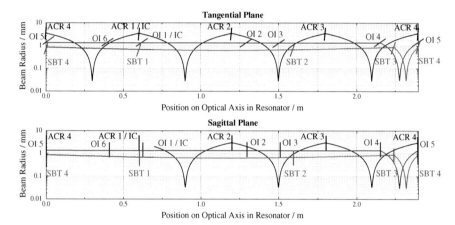

Fig. 5.2 The $1/e^2$-intensity radius along the optical axis of the three different cavity designs calculated with the ABCD matrix formalism for the tangential and sagittal planes, respectively. The mirror positions and AOI are indicated by markers drawn into the diagram. The laser spot sizes on the mirrors increase by the reciprocal cosine of the AOI. In the OI design the plotted beam dimension in the tangential plane has to be multiplied by a factor of $1/\cos 0.96 \approx 1.74$ for the plane and by $1/\cos 1.221 \approx 2.92$ for the parabolic mirrors. In the other two designs, this factor is almost 1

Fig. 5.1a[1] (the AOI in this example are $60°$). The second approach relies on detuning the resonator from the center of the stability range by means of curved mirror distance and AOI variation. In so doing, the irradiated areas on the mirrors can be increased and the sizes of intra-cavity beam waist radii can be decreased. While this can be carried out in any stable resonator, we exemplify this method with a symmetric all-curved-mirror resonator (ACR) design, where all mirrors have the same radius of curvature, see Fig. 5.1b. The main advantage of the symmetry is the fact that all spots on the mirrors are identical so that no mirror is exposed to an

---

[1] The graphic illustration was done by Jan Kaster.

intensity larger than any of the others. Further advantages of this resonator design include multiple foci (the central focus could be used for noncollinear HHG with two circulating pulses [2–4]), easy access to the foci for Thomson scattering experiments (see Sect. 5.4) and compactness. The resonator is detuned from the center of the stability range by means of the distances between the mirrors. In this way, the beam radius on the mirrors is increased and the focus size decreased. The limit is given by the beam ellipticity owing to the nonzero AOI on the mirrors, diffraction losses at the finite surfaces of the mirrors and mechanical stability. The third method relies on increasing the cavity length by an integer factor. The achieved spot size variation depends on the resonator design. In the case of the ACR, the spot areas on the mirrors will be increased by the square of the number of circulating pulses.

In the following, we compare the OI and the ACR designs to the widely used standard bow tie (SBT) design consisting of two spherically curved and at least two plane mirrors, as shown in Fig. 5.1c. For this comparison, we assume a central wavelength of 1,046 nm, a pulse duration of 250 fs, a pulse repetition rate of 250 MHz, two intra-cavity circulating pulses and fundamental-mode propagation with a beam waist radius of 30 μm. The pulse parameters were chosen according to the target parameters of a new laser system which is currently being built to seed the next-generation enhancement cavity in our group, see Sect. 5.1.2. Fixed beam waist radii correspond to certain positions in the resonator stability ranges and vice versa.

Figure 5.2 shows the beam radius along the cavity optical axis calculated with the ABCD matrix formalism. For the interpretation of the results of this comparison, it is essential to note that among the designs, several cavity parameters are not equivalent, such as e.g. the focal lengths f of the focusing mirrors.[2] The spot sizes on the mirrors are increased by a factor of 7.1 in the OI cavity (both focusing mirrors with $f = 120$ mm) and in the ACR cavity (all mirrors with $f = 150$ mm) by a factor of 14.1 compared to the SBT cavity (both focusing mirrors with $f = 75$ mm, as in the system presented in this thesis).

The next-generation LMA enhancement cavity system to be built in our group will have an ACR design. Currently this design is being experimentally implemented by Jan Kaster. The reasons for the choice of this design over the OI resonator include the larger achievable beam radii on the mirrors, simpler alignment and increased flexibility. Moreover, the currently available parabolic mirrors for an OI resonator exhibit significantly lower surface quality than the available spherical mirrors.

The experimental realization of an ACR means new challenges. On the one hand, in general large spot sizes imply an increased misalignment sensitivity (see e.g. [5]) and thus, require an increased mechanical stability. On the other hand, pushing the limitation related to nonlinear-effects will also allow for much higher average powers. Thus, absorption, scattering and thermal lensing in the cavity mirrors are expected to play an increasingly important role. In particular, thermal lensing increases the misalignment sensitivity (see e.g. [6]) and affects the transverse mode matching. As a

---

[2] For the OI cavity, f was chosen according to current manufacturing limitations for parabolic mirrors, for the SBT design f was chosen according to the system presented in this thesis and for the ACR design a convenient f was chosen.

measure towards solving these new problems and optimizing high-power operation, we plan to investigate the thermal conduction in the cavity mirrors theoretically and experimentally.

5.1.2 Novel Laser Sources

There are two directions in which further power scaling of the enhancement cavity technique is being pursued in our group. On the one hand, we aim at scaling up the average power of few-hundred-fs pulses. On the other hand, we pursue the reduction of the enhanced pulse duration. In the following we give a brief review of the experiments planned in these two directions, with emphasis on the laser sources.

5.1.2.1 Higher Average Power from Novel MHz Amplifier Systems

Among the available high-repetition-rate femtosecond laser systems, the most rapid growth of available average output power in the last few years has been achieved with Yb-based systems. Compared to the power available when this work was started, i.e. $\sim$50 W [7], the power available now is approximately one order of magnitude larger. The most prominent techniques allowing for this growth are fiber amplifiers [8] and the Innoslab concept [9, 10]. Initial power scaling experiments with the novel LMA cavity, currently being built, are planned with a modified version of the system presented in [8], currently under construction in the frame of a collaboration of our group with the fiber laser group at the IAP in Jena. The target parameters are 250 fs pulse duration with an average power of 500 W at a repetition rate of 250 MHz. For peak-intensity-related damage thresholds of state-of-the-art mirrors identified with our current setup, the ACR cavity in conjunction with the new seeding laser system offers the prospect of peak intensities on the order of 10^{15} W/cm^2 at intra-cavity beam waist radii of 30 μm. The circulating pulse duration is expected to remain unaffected and the circulating average power on the order of 1 MW. This regime is expected to boost the generated XUV power.

Besides highly efficient HHG, such a device with relatively long pulse durations would be ideally suited for the generation of hard X-ray radiation via inverse Compton (or Thomson) scattering, cf. Sect. 5.4. Here, the required laser pulse duration is lower bounded by the duration of the relativistic electron bunch and lies in the range of 1 ps.

5.1.2.2 Shorter Pulses Through Nonlinear Spectral Broadening

To obtain few-cycle femtosecond pulses, Ti:Sa-based laser systems are customarily employed. However, the challenges imposed by the short pulse duration on Ti:Sa amplifiers [11] have been limiting the achieved average powers to values significantly

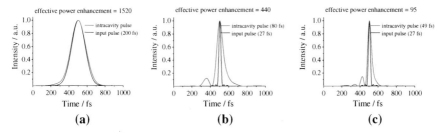

Fig. 5.3 Simulated pulse enhancement for transform-limited input pulses. The curves are normalized for clarity. **a** The conditions of some of the experiments presented in this thesis: 200 fs input pulse, input coupler with $R = 99.86\%$ and 7 highly reflecting mirrors. **b** Same mirrors, but 27 fs input pulse. The circulating pulse FWHM duration is 80 fs. **c** 27 fs input pulse and decreased finesse (with same GDD as before). The circulating pulse FWHM duration decreases to 49 fs. These simulations were carried out by Simon Holzberger

lower than those of longer pulses generated with Yb-based systems. Nonlinear spectral broadening and subsequent temporal compression can be used to shorten the inherently multi-cycle high-power pulses of Yb-based systems. For instance, the original layout of the laser system employed in the experiments presented in this thesis contains a nonlinear compression stage (which is currently bypassed), with which a pulse duration as short as 27 fs at an average output power exceeding 50 W was achieved at the repetition rate of 78 MHz [7]. In the meantime, even larger pulse energies have been obtained with the same technique, see e.g. [12], which were successfully used for direct, single-pass HHG. However, the process of spectral broadening in a fiber has both average and peak power limitations, which, in general, result in a longer pulse duration if the input parameters of the process are scaled up (e.g. 35 fs in [12]). Therefore, a very promising approach to highly efficient HHG consists of strong single-pass nonlinear compression with relaxed conditions on the input power and subsequent passive enhancement of the compressed pulses.

In our group Simon Holzberger works on the activation of the nonlinear pulse compression stage of the seeding system (see [7]) and its inclusion into the existing experimental setup. This implies additional major challenges. Firstly, the effect of the cavity single-round-trip dispersion on the circulating pulse in the steady state is magnified by a factor on the order of the power enhancement [13]. Moreover, dispersion also leads to a spectral filtering of the seeding comb. Thus, a bandwidth versus power enhancement trade-off needs to be found (cf. Sect. 2.1.2). To illustrate this, we have simulated the enhancement of the 200 fs pulses obtained before the nonlinear compression stage as well as the enhancement of the spectrally broadened and compressed pulses in the current cavity, using the calculated reflectivity and GDD curves for the mirrors provided by the manufacturer. Figure 5.3a depicts the enhancement of the 200 fs pulses. The calculated effective power enhancement (i.e. circulating average power divided by input average power) of 1,520 is in excellent agreement with the experimental results (see Chap. 4). The circulating pulse duration is not affected significantly. Figure 5.3b shows the result of using the same mirrors and seeding the cavity with the measured spectrum of the compressed 27-fs pulses if a flat spectral phase is assumed. We observe an effective power enhancement decrease to 440 and an intracavity pulse duration increase to 80 fs. Calculating with a simulated set of mirrors having the same dispersion properties but lower reflectivity (input coupler $R = 99.5\%$ and 7 cavity mirrors with $R = (99.5\%)^{(1/7)}$, i.e. providing an impedance matched peak enhancement of 200), yields the result depicted in Fig. 5.3c. Due to the reduced finesse and better impedance matching, a broader spectrum is coupled to the cavity, leading to a shorter intracavity pulse duration of 49 fs. However, the effective power enhancement is also reduced to 95. In conclusion, customized mirror designs with low *GDD* and/or the use of pairs of mirrors with opposite *GDD* are necessary. Research on an optimum mirror design is currently being performed in our group in collaboration with Vladimir Pervak and coworkers.

Amplitude and phase fluctuations stemming from the nonlinear spectral broadening process might hamper the resonant enhancement. To the best of our knowledge, spectrally broadened pulses have not yet been enhanced in a passive resonator. Additional stabilization precautions might become necessary, including the detection and stabilization of the second laser parameter, as suggested by the 50-nm cases in Fig. 2.9 in Sect. 2.1.3.

5.2 Power Scalable XUV Output Coupling

The first XUV generation experiments carried out with the system presented in this work are in good agreement with the theoretical predictions, see Sect. 4.5. The current limitation for experimentally investigating the HHG process in the new power regimes available in femtosecond enhancement cavities is given by the output coupling mechanism. In order to access the XUV radiation generated in these power regimes, novel XUV output coupling techniques need to be implemented first. In the course of this work, several new output couplers were developed and proposed, however, not yet tested. In the following, the techniques coming into consideration for the systems developed in our group will be discussed.

5.2.1 Thin Plates and WOMOC

A conceivable way to increase the damage threshold of thin plates, operated in transmission for the fundamental radiation and reflecting the XUV (such as Brewster plates or the GIP [14]), is improving their mounting with respect to cooling. In contrast to free-standing plates (see Fig. 4.11), the WOMOC is expected to allow a better cooling due to the fact that one surface of the wedged top layer, acting as the thin plate, is entirely attached to the underlying multilayer structure. To optimize cooling, the quantitative investigation of the heating effects is planned in our group. On the one hand, the in-situ measurement of the heating of these elements (and of the cavity mirrors) using a high-resolution thermal camera is planned. On the other hand, as for the cavity mirrors, we plan to develop a theoretical model for this process by numerically solving the heat conduction equation with appropriate boundary conditions.

5.2.2 Mirrors with Apertures, Direct Output Coupling

As mentioned in Sect. 4.4.3, mirrors with a circular hole with a suitable diameter have only very recently become available. Thus, to this day we have not carried out experiments with these novel mirrors. We expect that for a successful implementation of this method, additional precautions concerning the excited mode need to be taken. For instance, a position in the stability range will need to be chosen at which higher-order modes are not simultaneously resonant with the fundamental mode, thus avoiding a coupling among transverse modes. Additional spatial filtering with appropriate apertures might be necessary. Furthermore, the effect of the high intensity at the fundamental mode center on the hole edges is subject to future investigations. However, the major advantage of using the fundamental cavity mode is the already existing understanding of the HHG process for this transverse field distribution.

In parallel, quasi-imaging (QI) as presented in [15] is being investigated. To this end, curved mirrors with horizontal slits have been produced at the ILT in Aachen [16]. The manufacturing constraints of such slits are somewhat more relaxed than those of small holes, since the QI mode avoids the slit. In our cavity, the QI mode consisting of the degenerate Gauss-Hermite eigen-modes GH_{00} and GH_{40} experiences low losses even at slits with ~300 μm width. Initial power scaling results with such a mirror have revealed thermal limitations at around 5 kW of circulating power. The origin of this limitation is subject to further investigations. The production of new slotted mirrors with improved slit edges is planned in collaboration with the ILT. With those, further power scaling experiments and the investigation of thermal effects are planned. Moreover, in contrast to the fundamental mode, QI modes change their shape upon propagation along the optical axis due to the different Gouy phases of the involved resonator eigen-modes, cf. [15]. The relative rapid shape change in the region with an on-axis intensity maximum might affect the phase-matching of the HHG process. Furthermore, the spatial distribution of the generated high harmonics still needs to be investigated and the question whether the generated XUV light will pass through the slit needs to be answered.

Two further variations of the above methods are conceivable. Firstly, realizing QI in both transverse directions simultaneously could be achieved by canceling out the difference between the Gouy parameters for the two transverse directions, i.e. setting $\psi_x = \psi_y$. This could be done by horizontally tilting the beam incident on one of the curved mirrors in the current setup or by inserting an additional convex mirror [17]. In this case, a circular hole could be used instead of the slit. The rotational symmetric mode, which is a linear combination of Gauss-Laguerre eigen-modes in this case, might be beneficial for the HHG process. A second approach concerns the ACR design. To obtain relatively small foci, this resonator needs to be operated close to a stability range border. Therefore, QI in the center of the stability range cannot be achieved in this LMA cavity. However, as the stability range border is approached, the Gouy phase differences among higher-order modes decrease and the coupling among them increases. A quantitative description of this coupling is subject to further study.[3] We plan to investigate whether this coupling could be used to excite a QI-like mode in the ACR resonator with losses small enough so that they allow the desired power enhancement.

Finally, we mention that the currently implemented XUV detection setup (see Fig. 4.15) can easily be modified to suit the implementation of collinear, direct output coupling through a hole. The geometrically output coupled XUV beam can be steered to the XUV spectrometer by two parallel Au mirrors (or plates, anti-reflection coated for the IR [14]) under (close-to-) grazing incidence while the XUV photodiode mounted on the translation stage can remain at the same position.

5.3 Towards Isolated as-Pulses Using fs-Enhancement Cavities

In pump-probe experiments, probing with single pulses rather than a pulse train provides the best time resolution because the moment (or period) of observation is well-defined. In the frame of attosecond physics, the availability of isolated attosecond (as-) pulses at MHz repetition rates would enable new insights in the dynamics of the microcosm and is therefore highly desirable.

XUV as-pulses are generated with every linearly polarized optical half-cycle of the fundamental radiation driving the HHG process. Therefore, to obtain one isolated as-burst per driving fs pulse, either a single as-burst has to be isolated from the harmonic radiation generated by the driving pulse (which is usually a train of as-pulses), or the HHG process has to be confined to a single half-cycle of the driving pulse. The operation of methods following the first approach (e.g. *amplitude gating* [18]) is inherently limited to few-cycle pulses. Enhancing pulses this short in a passive resonator imposes extreme challenges to the mirror design. While the construction of such a setup would be a sensation, the second approach seems more realistic for obtaining isolated attosecond pulses from fs-enhancement cavities in the near future. In the following we give a brief overview of the most prominent techniques in line with this approach and discuss aspects relevant for implementing HHG gating in an enhancement cavity.

Polarization gating (PG) uses the fact that the HHG process is strongly polarization-dependent (see [19] and references therein). By combining two delayed counter-rotating circularly polarized pulses (with slightly different wavelengths), the polarization of the driving pulse is modulated in such a way that close-to linear

---

[3] Fig. 8a in the paper [15], see also Chap. 6, shows the power enhancement versus detuning from the stability range center. The width of this curve is given by the coupling among transverse modes, which increases again as the position in the stability range approaches an instability boundary.

polarization, necessary for HHG, is only achieved over a short time window within the resulting pulse, on the order of a half-cycle. The upper limit of the driving pulse duration is set by the ground state population depletion: if the pulse is too long, then the atoms will be fully ionized by the leading edge before the linear polarization gate starts and the as-burst can be generated. *Two-color gating* (TCG) employs waveforms synthesized from a fundamental-radiation pulse and its second harmonic (SH), which can increase the period between the generation of as-bursts to a full optical cycle of the fundamental, see e.g. [20]. Another technique relying on the few-cycle duration of the driving pulses is *ionization gating* (see e.g. [21] and references therein). Here, the first cycle reaching the intensity required for HHG generates an as-pulse and the subsequent cycles fully ionize the gas atoms so that HHG cannot occur anymore. A powerful method working also with multi-cycle driving pulses is *double optical gating* (DOG) [22], which combines PG and TCG. Finally, generalized double optical gating (GDOG) [23, 24] employs two elliptically (instead of circularly) polarized pulses to reduce the ground state population depletion at the leading edge of the pulse. GDOG allows the generation of isolated as-bursts with fundamental radiation pulses as long as 28 fs, generated by a Ti:Sa laser system. This value already lies in the pulse duration range for which enhancement in a passive cavity has been demonstrated [25]. Due to the larger wavelength, for an Yb-based system the same number of optical cycles corresponds to a pulse duration of approximately 36 fs, which can be achieved with nonlinear compression, cf. Sect. 5.1.2. Therefore, from the point of view of the pulse duration, the prospect of intracavity gating is given.

Another encouraging result is provided by a test we performed in our cavity. To verify the resonance of the cavity for elliptically polarized seeding light, we placed a quarter-wave plate in the input beam, just in front of the cavity input coupler. While we rotated the plate over 360°, the cavity scan pattern did not change observably, implying that the cavity losses for any ellipticity of the input polarization are comparable. Therefore, we assume that any linear combination of elliptically polarized pulses can be resonant in the cavity. This is a prerequisite for the majority of the gating methods mentioned here. The unique power regime achievable in enhancement cavities, which provides multi-cycle pulses with large intensities at MHz repetition rates, might also open the door to novel combinations of the existing techniques (e.g. a combination of PG with ionization gating seems promising) or to completely new gating mechanisms.

It should also be mentioned that superimposing a SH portion to the fundamental driving the HHG process can be readily implemented in a single-pass fashion. Standard highly reflecting dielectric mirrors usually have a transmission band at the second harmonic frequency of the fundamental light. Thus, the mirror just before the HHG focus could serve as the input coupler for the fundamental radiation, while a SH portion is transmitted through this mirror and overlaps coherently with the intracavity circulating pulse in the HHG focus (but is not resonant in the cavity).

Recently, Durach et al. [26, 27] predicted a plasmonic metallization of thin dielectric films, illuminated by intense single-cycle laser pulses. Such a film could be used as a reflective surface onto which the intracavity generated as-pulse train impinges. If a second pulse, with a slightly different wavelength (e.g. a Ti:Sa-generated) impinges on this surface, the metallization follows the electric field of the pulse virtually instantaneously and thus, could switch the reflectivity of the surface. This phenomenon could in principle be used as an alternative to the above methods for isolating an as-pulse from a train.

5.4 Other Experiments with High-Power Enhancement Cavities

An alternative way for converting the laser light to short wavelength radiation involves scattering the laser pulse from a relativistic electron pulse [28]. The relativistic Doppler shift leads to an increase of the photon energy by a factor $4\gamma^2$, where γ is the relativistic mass factor of the electrons. A 50 MeV electron beam thus generates 50 keV X-rays from 1.2 eV photons. Due to the small Thomson cross-section, a high laser pulse energy and high charge of the electron bunch are required for generating a useful number of X-ray photons. The weak nature of the interaction may be considered an advantage in case of an enhancement cavity since it induces negligible depletion of the circulating laser pulses [29]. We point out that the ACR cavity design shown in Fig. 5.1b is particularly suitable for Thomson scattering applications. Its geometry allows straightforward insertion of the electron pulse and extraction of the X-rays generated. If the electrons collide at a small angle with the laser pulse, only a small downshift of the X-ray photon energy is induced.

The high circulating power in the cavity combined with the high repetition rate affords applications of enhancement cavities in quite different fields of research. We mention multiphoton entanglement experiments [30], cavity-enhanced scattering [31], ultrasensitive absorption and dispersion spectroscopy [32], precision spectroscopy with helium [33], the generation of quantum frequency combs [34] and THz radiation generation [35]. For a successful realization of these applications a high circulating power is of crucial importance.

References

1. J. Kaster, I. Pupeza, T. Eidam, C. Jocher, E. Fill, J. Limpert, R. Holzwarth, B. Bernhardt, T. Udem, T.W. Hänsch, A. Tünnermann, F. Krausz, Towards MW average powers in ultrafast high-repetition-rate enhancement cavities. HILAS, paper HFB4 (2011)
2. S.V. Fomichev, P. Breger, B. Carre, P. Agostini, D.F. Zaretski, Non-collinear high harmonic generation. Laser Phys. **12**, 383 (2002)
3. A. Ozawa, A. Vernaleken, W. Schneider, I. Gotlibovych, T. Udem, T.W. Hänsch, Non-collinear high harmonic generation: a promising outcoupling method for cavity-assisted XUV generation. Opt. Express **16**, 6233 (2008)
4. K. Moll, R. Jones, J. Ye, Output coupling methods for cavity-based high-harmonic generation. Opt. Express **14**, 8189 (2006)
5. R. Hauck, H.P. Kortz, H. Weber, Misalignment sensitivity of optical resonators. Appl. Optics **19**, 598 (1980)

6. V. Magni, Resonators for solid-state lasers with large-volume fundamental mode and high alignment stability. Appl. Optics **25**, 107 (1986)
7. T. Eidam, F. Röser, O. Schmidt, J. Limpert, A. Tünnermann, 57 W, 27 fs pulses from a fiber laser system using nonlinear compression. Appl. Phys. B **92**, 9 (2008)
8. T. Eidam, S. Hanf, E. Seise, T.V. Andersen, T. Gabler, C. Wirth, T. Schreiber, J. Limpert, A. Tünnermann, Femtosecond fiber CPA system emitting 830 W average output power. Opt. Lett. **35**, 94 (2010)
9. P. Russbüldt, T. Mans, G. Rotarius, J. Weitenberg, H.D. Hoffmann, R. Poprawe, 400 W Yb:YAG Innoslab fs-Amplifier. Opt. Express **17**, 12230 (2009)
10. P. Russbüldt, T. Mans, J. Weitenberg, H.D. Hoffmann, R. Poprawe, Compact diode-pumped 1.1 kW Yb:YAG Innoslab femtosecond amplifier. Opt. Lett. **35**, 4169 (2010)
11. A. Ozawa, T. Udem, U.D. Zeitner, T.W. Hänsch, P. Hommelhoff, Modeling and optimization of single-pass laser amplifiers for high-repetition-rate laser pulses. Phys. Rev. A **82**, 033815 (2010)
12. A. Vernaleken, J. Weitenberg, T. Sartorius, P. Russbüldt, W. Schneider, S.L. Stebbings, M.F. Kling, P. Hommelhoff, H.-D. Hoffmann, R. Poprawe, F. Krausz, T.W. Hänsch, T. Udem, Single pass high harmonic generation at 20.8 MHz repetition rate. Opt. Lett. **36**, 3428 (2011)
13. I. Pupeza, X. Gu, E. Fill, T. Eidam, J. Limpert, A. Tünnermann, F. Krausz, T. Udem, Highly sensitive dispersion measurement of a high-power passive optical resonator using spatial-spectral interferometry. Opt. Express **18**, 26184 (2010)
14. O. Pronin, V. Pervak, E. Fill, J. Rauschenberger, F. Krausz, A. Apolonski, Ultrabroadband efficient intracavity XUV output coupler. Opt. Express **19**, 10232 (2011)
15. J. Weitenberg, P. Russbüldt, T. Eidam, I. Pupeza, Transverse mode tailoring in a quasi-imaging high-finesse femtosecond enhancement cavity. Opt. Express **19**, 9551 (2011)
16. D. Esser, W. Bröring, J. Weitenberg, H.-D. Hoffmann, Laser-manufactured mirrors for geometrical output coupling of intracavity-generated high harmonics. manuscript in preparation
17. J. Weitenberg, personal communication
18. E. Goulielmakis, M. Schultze, M. Hofstetter, V.S. Yakovlev, J.G.M. Uiberacker, A.L. Aquila, E.M. Gullikson, D.T. Attwood, R. Kienberger, F. Krausz, U. Kleineberg, Single-Cycle nonlinear optics. Science **320**, 1614 (2008)
19. I.J. Sola, E. Mevel, L. Elouga, E. Constant, V. Strelkov, L. Poletto, P. Villoresi, E. Benedetti, J.-P. Caumes, S. Stagira, C. Vozzi, G. Sansone, M. Nisoli, Controlling attosecond electron dynamics by phase-stabilized polarization gating. Nat. Phys. **2**, 319 (2006)
20. M. Fiess, B. Horvath, T. Wittmann, W. Helml, Y. Cheng, B. Zeng, Z. Xu, A. Scrinzi, J. Gagnon, F. Krausz, R. Kienberger, Attosecond control of tunneling ionization and electron trajectories. New J. Phys. **13**, 033031 (2011)
21. M.J. Abel, T. Pfeifer, P.M. Nagel, W. Boutu, M.J. Bell, C.P. Steiner, D.M. Neumark, S.R. Leone, Isolated attosecond pulses from ionization gating of high-harmonic emission. Chem. Phys. **366**, 9 (2009)
22. H. Mashiko, S. Gilbertson, C. Li, S.D. Khan, M.M. Shakya, E. Moon, Z. Chang, Double optical gating of high-order harmonic generation with carrier-envelope phase stabilized lasers. Phys. Rev. Lett. **100**, 103906 (2008)
23. X. Feng, S. Gilbertson, H. Mashiko, H. Wang, S.D. Khan, M. Chini, Y. Wu, K. Zhao, Z. Chang, Generation of isolated attosecond pulses with 20 to 28 femtosecond lasers. Phys. Rev. Lett. **103**, 183901 (2009)
24. S. Gilbertson, Y. Wu, S.D. Khan, M. Chini, K. Zhao, X. Feng, Z. Chang, Isolated attosecond pulse generation using multicycle pulses directly from a laser amplifier. Phys. Rev. A **81**, 043810 (2010)
25. C. Gohle, T. Udem, M. Herrmann, J. Rauschenberger, R. Holzwarth, H.A. Schuessler, F. Krausz, T.W. Hänsch, A frequency comb in the extreme ultraviolet. Nature **436**, 234 (2005)
26. M. Durach, A. Rusina, M.F. Kling, M.I. Stockman, Metallization of nanofilms in strong adiabatic electric fields. Phys. Rev. Lett. **105**, 086 803 (2010)
27. M. Durach, A. Rusina, M.F. Kling, M.I. Stockman, Ultrafast dynamic matallization of dielectric nanofilms by strong single-cycle optical fields. submitted for publication (2011)

28. F.V. Hartemann, W.J. Brown, D.J. Gibson, S.G. Anderson, A.M. Tremaine, P.T. Springer, A.J. Wootton, E.P. Hartouni, C.P.J. Barty, High-energy scaling of Compton scattering light sources. Phys. Rev. ST Accel. Beams **8**, 100702 (2005)

29. K. Sakaue, M. Washio, S. Araki, M. Fukuda, Y. Higashi, Y. Honda, T. Omori, T. Taniguchi, N. Terunum, J. Urakawa, N. Sasao, Observation of pulsed X-ray trains produced by laser-electron compton scatterings. Rev. Sci. Instrum. **80**, 123304 (2009)

30. R. Krischek, W. Wieczorek, A. Ozawa, N. Kiesel, P. Michelberger, T. Udem, H. Weinfurter, Ultraviolet enhancement cavity for ultrafast nonlinear optics and high-rate multiphoton entanglement experiments. Nat. Phot. **4**, 170 (2010)

31. M. Motsch, M. Zeppenfeld, P.W.H. Pinkse, G. Rempe, Cavity-enhanced rayleigh scattering. New J. Phys. **12**, 063022 (2010)

32. C. Gohle, B. Stein, A. Schliesser, T. Udem, T.W. Hänsch, Frequency comb vernier spectroscopy for broadband, high resolution, high sensitivity absorption and dispersion spectra. Phys. Rev. Lett. **99**, 263902 (2007)

33. M. Herrmann, M. Haas, U.D. Jentschura, F. Kottmann, D. Leibfried, G. Saathoff, C. Gohle, A. Ozawa, V. Batteiger, S. Knünz, N. Kolachevsky, H.A. Schüssler, T.W. Hänsch, T. Udem, Feasibility of coherent xuv spectroscopy on the $1S$–$2S$ transition in singly ionized helium. Phys. Rev. A **79**, 052505 (2009)

34. A. Zavatta, V. Parigi, M. Bellini, Towards quantum frequency combs: Boosting the generation of highly nonclassical light states by cavity-enhanced parametric down-conversion at high repetition rates. Phys. Rev. A **78**, 033809 (2008)

35. M. Theuer, D. Molter, K. Maki, C. Otani, J.A. L'huillier, R. Beigang, Terahertz generation in an actively controlled femtosecond enhancement cavity. Appl. Phys. Lett. **93**, 041119 (2008)

Chapter 6
Results Published in the Course of this Project

In the course of this doctorate project, several results obtained by the author have been published in peer-reviewed journals. This chapter summarizes these results by quoting the abstracts of the respective articles.

I. Pupeza, T. Eidam, J. Rauschenberger, B. Bernhardt, A. Ozawa, E. Fill, A. Apolonski, Th. Udem, J. Limpert, Z. A. Alahmed, A. M. Azzeer, A. Tünnermann, T. W. Hänsch, F. Krausz, "Power scaling of a high repetition rate enhancement cavity," Opt. Letters 12, 2052–2054 (2010).

Abstract. A passive optical resonator is used to enhance the power of a pulsed 78 MHz repetition rate Yb laser providing 200 fs pulses. We find limitations relating to the achievable time-averaged and peak power which we distinguish by varying the duration of the input pulses. An intra-cavity average power of 18 kW is generated with close to Fourier-limited pulses of 10 W average power. Beyond this power level, intensity-related effects lead to resonator instabilities which can be removed by chirping the seed laser pulses. Extending the pulse duration in this way to 2 ps we could obtain 72 kW of intra-cavity circulating power with 50 W of input power.

I. Pupeza, X. Gu, E. Fill, T. Eidam, J. Limpert, A. Tünnermann, F. Krausz, Th. Udem, "Highly sensitive dispersion measurement of a high-power passive optical resonator using spatial-spectral interferometry," Opt. Express 18, 26184–26195 (2010).

Abstract. We apply spatially and spectrally resolved interferometry to measure the complex ratio between the field circulating inside a high-finesse femtosecond enhancement cavity and the seeding field. Our simple and highly sensitive method enables the measurement of single-round-trip group delay dispersion of a fully loaded cavity at resonance for the first time. Group delay dispersion can be determined with a reproducibility better than $1\,\mathrm{fs}^2$ allowing the investigation of nonlinear processes triggered by the high intracavity power. The required data acquisition time is less than 1 s.

I. Pupeza, E. Fill, F. Krausz, "Low-loss VIS/IR-XUV beam splitter for high-power applications," Opt. Express 19, 12108–12118 (2011).

I. Pupeza, *Power Scaling of Enhancement Cavities for Nonlinear Optics*, 85
Springer Theses, DOI: 10.1007/978-1-4614-4100-7_6,
© Springer Science+Business Media New York 2012

Abstract. We present a low-loss VIS/IR-XUV beam splitter, suitable for high-power operation. The spatial separation of the VIS/IR and XUV components of a beam is achieved by the wedged top layer of a dielectric multilayer structure, onto which the beam is impinging under Brewster's angle (for VIS/IR). With a fused silica wedge with an angle of $0.5°$ we achieve a separation angle of $2.2°$ and an IR reflectivity of 0.9995. Typical XUV reflectivities amount to $0.1–0.2$. The novel element is mechanically robust, exhibiting two major advantages over free-standing Brewster plates: (i) a significant improvement of heat conduction and (ii) easier handling, in particular for high-optical-quality fabrication. The beam splitter could be used as an output coupler for intracavity-generated XUV radiation, promising a boost of the power regime of current MHz-HHG experiments. It is also suited for single-pass experiments and as a beam combiner for pump-probe experiments.

Y.-Y. Yang, F. Süßmann, S. Zherebtsov, I. Pupeza, J. Kaster, D. Lehr, H.-J. Fuchs, E.-B. Kley, E. Fill, X.-M. Duan, Z.-S. Zhao, F. Krausz, S. L. Stebbings, M. F. Kling, "Optimization and characterization of a highly-efficient diffraction nanograting for MHz XUV pulses," Opt. Express 19, 1955–1962 (2011).

Abstract. We designed, fabricated and characterized a nano-periodical highly-efficient blazed grating for extreme-ultraviolet (XUV) radiation. The grating was optimized by the rigorous coupled-wave analysis method (RCWA) and milled into the top layer of a highly-reflective mirror for IR light. The XUV diffraction efficiency was determined to be around 20% in the range from 35.5 to 79.2 nm. The effects of the nanograting on the reflectivity of the IR light and non-linear effects introduced by the nanograting have been measured and are discussed.

J. Weitenberg, P. Rußbüldt, T. Eidam, I. Pupeza, "Transverse mode tailoring in a quasi-imaging high-finesse femtosecond enhancement cavity," Opt. Express 19, 9551–9561 (2011).

Abstract. We demonstrate a high-finesse femtosecond enhancement cavity with an on-axis obstacle. By inserting a wire with a width of 5% of the fundamental mode diameter, the finesse of $F = 3400$ is only slightly reduced to $F = 3000$. The low loss is due to the degeneracy of transverse modes, which allows for exciting a circulating field distribution avoiding the obstacle. We call this condition quasi-imaging. The concept could be used for output coupling of intracavity-generated higher-order harmonics through an on-axis opening in one of the cavity mirrors.

Chapter 7
Appendix

7.1 Experiments Involving Enhancement Cavities

Continuous-wave passive cavities with power enhancement factors on the order of 10^5 have been demonstrated in the early nineties for sensitivity enhancement of optical losses measurements, see e.g. [1]. With the advent of Ti:Sa-based frequency combs, ultrashort pulses were enhanced for the first time for MHz-repetition-rate HHG [2, 3]. Due to their broad optical bandwidth, Ti:Sa lasers enable high peak intensities through the short pulse duration rather than the average power. The shortest intra-cavity pulse duration of 27 fs has been achieved with a power enhancement of 50 by seeding a broadband cavity with 20 fs pulses [2]. More recently, Yb-based frequency combs were employed for intra-cavity HHG due to their significantly larger average power [4, 5]. So far, the largest circulating powers in an ultrafast enhancement cavity lie in the multi-10-kW range [6] and have been demonstrated with the Yb-based system presented in this thesis. For some applications, the only drawback of Yb-based seeding lasers towards Ti:Sa systems is the relatively narrow available optical bandwidth. However, nonlinear pulse compression of the seeding pulses offers the prospect of the enhancement of few-cycle Yb-based pulses (see also Sect. 5.1.2).

Further passively enhanced laser sources include frequency-doubled Ti:Sa and Yb-based systems in the UV [7] and in the green spectral regions [8], respectively. However, the available mirrors for these wavelengths reach significantly lower reflectivity values than in the infrared, limiting the achievable finesse. The UV enhancement cavity, which also includes a nonlinear crystal, supports 7 W of circulating average power at a repetition rate of 81 MHz enabling a powerful source for experiments with entangled multiphoton states. The green enhancement cavity is an ongoing project led by Birgitta Bernhardt in the group of Prof. Hänsch at the MPQ, in collaboration with our group. The goal of this experiment is to increase the available power of frequency combs at around 60 nm wavelength for high-precision XUV spectroscopy. Using green light for HHG promises a significant increase in

I. Pupeza, *Power Scaling of Enhancement Cavities for Nonlinear Optics*,
Springer Theses, DOI: 10.1007/978-1-4614-4100-7_7,
© Springer Science+Business Media New York 2012

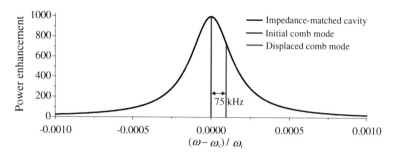

Fig. 7.1 Enhancement of mode number 3695666 (*blue line*) of a 78-MHz repetition-rate frequency comb with the central wavelength of 1040 nm in an impedance-matched enhancement cavity with a power enhancement factor of 1,000. A frequency shift $\Delta f = 7.5$ kHz of the enhanced mode (*red line*) leads to an enhancement reduction to $\sim$725. This comb shift corresponds to either a repetition frequency change of 2 mHz (equivalent to a roundtrip length change of 0.1 nm) or to a CE phase slippage of 0.6 mrad

the conversion efficiency due to the shorter fundamental wavelength, see Sect. 2.2.2. So far, the circulating power in the cavity is limited to a few hundred watts due to a mirror degradation phenomenon in vacuum, which is still being investigated [8, 9].

7.2 An Example of Comb-Cavity Detuning

To exemplify the effect of a detuning of the cavity length on the enhancement, we consider the case of a 78-MHz repetition-rate frequency comb, with the central optical wavelength of 1,040 nm. These values are chosen in accordance with the system presented in this thesis. Furthermore, we consider an impedance-matched enhancement cavity with a power enhancement factor of 1,000, so that the central comb mode (with wavelength 1,040 nm and mode number 3695666) is locked to a corresponding cavity resonance, see blue line in Fig. 7.1. An optical frequency shift of the enhanced comb line of $\Delta f = 7.5$ kHz means a reduction of the enhancement to $\sim$725, see red line in Fig. 7.1. This shift corresponds to either a repetition frequency change of $\Delta f / 3695666 = 2$ mHz of the seeding comb, which is equivalent to a roundtrip length change of 0.1 nm or to a CE phase slippage of $\Delta \varphi = 2\pi \cdot 7.5 \cdot 10^3 / (78 \cdot 10^6)$ rad$= 0.6$ mrad.

7.3 Overview of XUV Output Coupling Methods

In the following we give a brief overview of the existing methods for coupling out intracavity generated XUV radiation, i.e. spatially splitting the generated high harmonics (HH) from the circulating fundamental radiation (FR).

Brewster Plates

- *Mode of operation:* the HH experience total internal reflection at a thin plate placed under Brewster's angle for the FR.
- *Publications:* [2, 3, 10]; *discussed in this thesis in* Sects. 4.4.1 *and* 5.2.1.
- *Experimentally demonstrated:* yes.
- *Output coupling efficiency at* 60 nm: roughly 10%, *at* 10 nm: $\ll 0.1\%$.
- *Average power scalability:* poor due to poor heat conduction. Currently limited at roughly 3 kW. *Peak power scalability:* poor due to nonlinearities which increase with thickness, (but can in principle be compensated for).

GIP: Grazing-Incidence Plates

- *Mode of operation:* similar to the Brewster plate, but under grazing incidence and (optionally) for s polarization. High transmission for the FR through the GIP is obtained by anti-reflection coatings on both sides of the plate.
- *Publications:* [11]; *not further discussed in this thesis.*
- *Experimental demonstration:* has not been included in a cavity yet.
- *Output coupling efficiency at* 60 nm: roughly 60%, *at* 10 nm: roughly 5%.
- *Average and peak power scalability:* not yet investigated. Expected to be similar to the Brewster plate, or poorer due to the additional coatings.
- *Additional nonlinearities and dispersion:* similar to the Brewster plate.

WOMOC: Wedge-on-Mirror Output Coupler

- *Mode of operation:* a thin plate is attached to the surface of a rigid, highly reflecting mirror, providing improved thermal conduction. The FR impinges on the WOMOC under Brewster's angle, penetrates the plate, is reflected by the mirror and exits the plate. The plate is wedged under a small angle, ensuring the spatial separation of the HH from the FR.
- *Publications:* [12]; *discussed in this thesis in* Sects. 4.4.2 *and* 5.2.1.
- *Experimental demonstration:* has not been included in a cavity yet.
- *Output coupling efficiency at* 60 nm *and at* 10 nm: like Brewster plate.
- *Average and peak power scalability:* not yet investigated. Expected to be better than Brewster plate because (i) the penetrated material is attached over a large area to a robust optic and (ii) the penetrated material thickness is decreased.
- *Additional nonlinearities and dispersion:* expected to be low or negligible.

Nanograting

- *Mode of operation:* a nanostructure written in the last layer of the dielectric multi-layer coating of the cavity mirror following the HHG focus diffracts the HH away from the resonator optical axis. For the FR the element acts as a high reflector.
- *Publications:* [4, 13]; *discussed in this thesis in* Sect. 4.4.4.
- *Experimental demonstration:* yes.
- *Output coupling efficiency at 60 nm and at 10 nm:* similar to Brewster plate or better. Can be optimized for a specific wavelength region.
- *Average power scalability:* expected to be very good. *Peak power scalability:* this is a limitation due to the nonlinearities enhanced by the nanostructure.
- *Additional nonlinearities:* not known. *Additional dispersion:* none to the IR. However, the output coupled harmonics are spatially dispersed.

Hole in Mirror After HHG Focus and Use of Fundamental Mode

- *Mode of operation:* the HH, whose divergence decreases with increasing order, are directly coupled out through a small on-axis hole in the cavity mirror following the HHG focus. If the hole is small enough, the losses introduced to the circulating FR still allow for a high enhancement.
- *Publications:* [14, 15]; *discussed in this thesis in* Sect. 5.2.2.
- *Experimental demonstration:* while a cavity with such a mirror has been demonstrated, output coupling of high harmonics through the hole has not yet been reported.
- *Output coupling efficiency at 60 nm and at 10 nm:* depends on the directional characteristic of the XUV radiation. It is expected to increase for lower wavelengths.
- *Average power scalability:* not yet investigated. It is expected to depend critically on the quality of the hole edge. *Peak power scalability:* not yet investigated. It is expected to be good.
- *Additional nonlinearities and dispersion:* none.

Hole or Slit in Mirror After HHG Focus and Quasi-Imaging

- *Mode of operation:* using quasi-imaging, for the FR a field distribution is excited in the cavity, which has an on-axis maximum near the focus for HHG and simultaneously avoids a hole or slit in the cavity mirror following the focus. The HH generated in the central lobe of the on-axis maximum are expected to propagate along the axis and exit the cavity through the opening in the mirror.
- *Publications:* [16, 17]; *discussed in this thesis in* Sects. 2.1.4, 4.4.3 *and* 5.2.2.
- *Experimental demonstration:* a proof of principle of quasi-imaging was demonstrated. However, the suitability of quasi-imaging modes for HHG as well as output coupling of XUV light through an opening in a mirror following the focus has not yet been shown.

- *Output coupling efficiency at 60 nm and at 10 nm:* depends on the directional characteristic of the generated XUV radiation, which still needs to be investigated. It is expected to be very high for lower wavelengths.
- *Average power scalability:* not investigated properly yet. Preliminary results in our lab show 5 kW. It is expected to depend on the quality of the slit edge. *Peak power scalability:* not investigated yet. It is expected to be good.
- *Additional nonlinearities and dispersion:* none.

Non-Collinear HHG

- *Mode of operation:* two FR pulses collide non-collinearly in a high-intensity region of one (or two) cavity/cavities, creating an interference pattern. For properly chosen parameters, the emitted HH propagate along the bisector of the angle formed by the two propagation paths of the FR.
- *Publications:* [14, 18, 19]; *not further discussed in this thesis.*
- *Experimental demonstration:* a proof of principle for the single-pass non-collinear HHG has been shown. However, this technique has not yet been shown in conjunction with enhancement cavities.
- *Output coupling efficiency at 60 nm and at 10 nm:* not investigated yet but expected to be very good.
- *Average and peak power scalability:* not investigated yet but expected to perform very well.
- *Additional nonlinearities and dispersion:* none.

References

1. G. Rempe, R.J. Thompson, H.J. Kimble, R. Lalezari, Measurement of ultralow losses in an optical interferometer. Opt. Lett. **17**, 363 (1992)
2. C. Gohle, T. Udem, M. Herrmann, J. Rauschenberger, R. Holzwarth, H.A. Schuessler, F. Krausz, T.W. Hänsch, A frequency comb in the extreme ultraviolet. Nature **436**, 234 (2005)
3. R. Jones, K.D. Moll, M.J. Thorpe, J. Ye, Phase-Coherent Frequency Combs in the Vacuum Ultraviolet via high-harmonic generation inside a femtosecond enhancement cavity. Phys. Rev. Lett. **94**, 193201 (2005)
4. D.C. Yost, T.R. Schibli, J. Ye, Efficient output coupling of intracavity high harmonic generation. Opt. Lett. **33**, 1099–1101 (2008)
5. A. Cingöz, D.C. Yost, J. Ye, A. Ruehl, M. Fermann, I. Hartl, Power scaling of high-repetition-rate HHG. International Conference on Ultrafast Phenomena, 2010
6. I. Pupeza, T. Eidam, J. Rauschenberger, B. Bernhardt, A. Ozawa, E. Fill, A. Apolonski, T. Udem, J. Limpert, Z.A. Alahmed, A.M. Azzeer, A. Tünnermann, T.W. Hänsch, F. Krausz, Power scaling of a high repetition rate enhancement cavity. Opt. Lett. **12**, 2052 (2010)
7. R. Krischek, W. Wieczorek, A. Ozawa, N. Kiesel, P. Michelberger, T. Udem, H. Weinfurter, Ultraviolet enhancement cavity for ultrafast nonlinear optics and high-rate multiphoton entanglement experiments. Nat. Phot. **4**, 170 (2010)

8. B. Bernhardt, A. Ozawa, I. Pupeza, A. Vernaleken, Y. Kobayashi, R. Holzwarth, E. Fill, F. Krausz, T.W. Hänsch, T. Udem, Green enhancement cavity for frequency comb generation in the extreme ultraviolet. CLEO, paper QTuF3, 2011
9. B. Bernhardt, personal communication
10. A. Ozawa, J. Rauschenberger, C. Gohle, M. Herrmann, D.R. Walker, V. Pervak, A. Fernandez, R. Graf, A. Apolonski, R. Holzwarth, F. Krausz, T. Hänsch, T. Udem, High harmonic frequency combs for high resolution spectroscopy. Phys. Rev. Lett. **100**, 253901 (2008)
11. O. Pronin, V. Pervak, E. Fill, J. Rauschenberger, F. Krausz, A. Apolonski, Ultrabroadband efficient intracavity XUV output coupler. Opt. Express **19**, 10232 (2011)
12. I. Pupeza, E. Fill, F. Krausz, Low-loss VIS/IR-XUV beam splitter for high-power applications. Opt. Express **19**, 12108 (2011)
13. Y.-Y. Yang, F. Süssmann, S. Zherebtsov, I. Pupeza, J. Kaster, D. Lehr, E.-B. Kley, E. Fill, X.-M. Duan, Z.-S. Zhao, F. Krausz, S. Stebbings, M.F. Kling, Optimization and characterization of a highly-efficient diffraction nanograting for MHz XUV pulses. Opt. Express **19**, 1955 (2011)
14. K. Moll, R. Jones, J. Ye, Output coupling methods for cavity-based high-harmonic generation. Opt. Express **14**, 8189 (2006)
15. D. Esser, W. Bröring, J. Weitenberg, H.-D. Hoffmann, Laser-manufactured mirrors for geometrical output coupling of intracavity-generated high harmonics, manuscript in preparation
16. J. Weitenberg, P. Russbüldt, T. Eidam, I. Pupeza, Transverse mode tailoring in a quasi-imaging high-finesse femtosecond enhancement cavity. Opt. Express **19**, 9551 (2011)
17. J. Weitenberg, P. Russbüldt, I. Pupeza, T. Udem, H.-D. Hoffmann, and R. Poprawe, "Geometrical on-axis access to high-finesse resonators by quasi-imaging", manuscript in preparation
18. S.V. Fomichev, P. Breger, B. Carre, P. Agostini, D.F. Zaretski, Non-collinear high harmonic generation. Laser Phys. **12**, 383 (2002)
19. A. Ozawa, A. Vernaleken, W. Schneider, I. Gotlibovych, T. Udem, T.W. Hänsch, Non-collinear high harmonic generation: a promising outcoupling method for cavity-assisted XUV generation. Opt. Express **16**, 6233 (2008)

Curriculum Vitae

Date of birth   9 March 1980
Place of birth   Bucharest, Romania
Citizenship   German

Education

1986–1997   Deutsche Schule Bukarest
1997–1999   Scharnhorstgymnasiums Hildesheim, Abitur
1999–2006   Studies of Electrical Engineering at the Technical University Braunschweig
2000–2007   Studies of Mathematics at the Technical University Braunschweig
2004–2005   Visiting fellow and diploma thesis in Electrical Engineering at the Harvard University, Cambridge, MA, USA with a stipend of the German National Academic Foundation
2005   Internship at the German Aerospace Center (DLR) in Oberpfaffenhofen
2007–2011   Ph.D. student in the group of Prof. Ferenc Krausz at the Max Planck Institute for Quantum Optics in Garching

Publications 2007–2011

Journal Publications

- I. Pupeza, R. Wilk, M. Koch, "Highly accurate optical material parameter determination with THz time domain spectroscopy," Opt. Express **15**, 4335–4350 (2007).

- R. Wilk, I. Pupeza, R. Cernat, M. Koch, "Highly Accurate THz Time-Domain Spectroscopy of Multi-Layer Structures," IEEE Journal of Selected Topics in Quantum Electronics **14**, 391–398 (2008).
- T. Kleine-Ostmann, T. Schrader, M. Bieler, U. Siegner, C. Monte, B. Gutschwager, J. Hollandt, A. Steiger, L. Werner, R. Müller, G. Ulm, I. Pupeza, M. Koch, "THz Metrology," Frequenz **62**, 137–148 (2008).
- M. A. Salhi, I. Pupeza, M. Koch, "Confocal THz Microscope," Journal of Infrared, Millimeter and Therahertz Waves **31**, 358–366 (2009).
- C. Jördens, K. L. Chee, I. A. I. Al-Naib, I. Pupeza, S. Peik, G. Wenke, M. Koch, "Dielectric Fibres for Low-Loss Transmission of Millimetre Waves and its Application in Couplers and Splitters," Journal of Infrared, Millimeter and Therahertz Waves **31**, 214–220 (2009).
- N. Krumbholz, T. Hochrein, N. Vieweg, I. Radovanovic, I. Pupeza, M. Schubert, K. Kretschmer, M. Koch, "Degree of Dispersion of Polymeric Compounds Determined with Terahertz Time-Domain Spectroscopy," Polymer Engineering and Science **51**, 109–116 (2010).
- S. Immervoll, R. Löwen, I. Pupeza, "A local characterization of smooth projective planes," Proceedings of the American Mathematical Society **138**, 323–332 (2010).
- I. Pupeza, T. Eidam, J. Rauschenberger, B. Bernhardt, A. Ozawa, E. Fill, A. Apolonski, Th. Udem, J. Limpert, Z. A. Alahmed, A. M. Azzeer, A. Tünnermann, T. W. Hänsch and F. Krausz, "Power scaling of a high repetition rate enhancement cavity," Opt. Letters **12**, 2052–2054 (2010).
- I. Pupeza, X. Gu, E. Fill, T. Eidam, J. Limpert, A. Tünnermann, F. Krausz, Th. Udem, "Highly sensitive dispersion measurement of a high-power passive optical resonator using spatial-spectral interferometry," Opt. Express **18**, 26184–26195 (2010).
- I. Pupeza, T. Eidam, J. Kaster, B. Bernhardt, J. Rauschenberger, A. Ozawa, E. Fill, Th. Udem, M. F. Kling, J. Limpert, Z. A. Alahmed, A. M. Azzeer, A. Tünnermann, Th. W. Hänsch, F. Krausz, "Power scaling of femtosecond enhancement cavities and high-power applications," Proc. SPIE **7914**, Paper 79141I (2011).
- Y.-Y. Yang, F. Süßmann, S. Zherebtsov, I. Pupeza, J. Kaster, D. Lehr, H.-J. Fuchs, E.-B. Kley, E. Fill, X.-M. Duan, Z.-S. Zhao, F. Krausz, S. L. Stebbings, M. F. Kling, "Optimization and characterization of a highly-efficient diffraction nanograting for MHz XUV pulses," Opt. Express **19**, 1955–1962 (2011).
- J. Weitenberg, P. Rußbüldt, T. Eidam, I. Pupeza, "Transverse mode tailoring in a quasi-imaging high-nesse femtosecond enhancement cavity," Opt. Express **19**, 9551–9561 (2011).
- I. Pupeza, E. Fill, F. Krausz, "Low-loss VIS/IR-XUV beam splitter for high-power applications" Opt. Express **19**, 12108–12118 (2011).
- J. Weitenberg, P. Rußbüldt, I. Pupeza, Th. Udem, H.-D. Homann, an R. Poprawe, "Geometrical on-axis access to high-finesse resonators by quasi-imaging," manuscript in preparation.

- B. Bernhardt, A. Ozawa, A. Vernaleken, I. Pupeza, J. Kaster, Y. Kobayashi, R. Holzwarth, E. Fill, F. Krausz, T. W. Hänsch, and Th. Udem, "Vacuum Ultraviolet Frequency Combs Generated by a Femtosecond Enhancement Cavity in the Visible," Opt. Letters **37**, 503–505 (2012).

Contributed and Invited Conference Talks (Selection)

- I. Pupeza, R. Wilk, F. Rutz, M. Koch, "Highly Accurate Material Parameter Extraction from THz Time Domain Spectroscopy Data," CLEO (2007).
- I. Pupeza, J. Rauschenberger, T. Eidam, F. Röser, B. Bernhardt, A. Ozawa, R. Holzwarth, Th. Udem, J. Limpert, A. Apolonski, T. W. Hänsch, A. Tünnermann, F. Krausz, "Towards High-Power XUV Generation Using an Yb-Based Enhancement Cavity," CLEO/Europe (2009).
- I. Pupeza, T. Eidam, B. Bernhardt, A. Ozawa, O. Pronin, J. Rauschenberger, F. Röser, A. Apolonski, Th. Udem, J. Limpert, A. Tünnermann, T. W. Hänsch, F. Krausz, "High-Power Femtosecond Enhancement Cavities for XUV Generation," invited talk, International Laser Physics Workshop (2009).
- I. Pupeza, T. Eidam, O. Pronin, J. Rauschenberger, B. Bernhardt, A. Ozawa, Th. Udem, R. Holzwarth, J. Limpert, A. Apolonski, T. W. Hänsch, A. Tünnermann, F. Krausz, "Femtosecond High Repetition Rate External Cavity beyond the Average Power Limit for Linear Enhancement," ASSP (2010).
- I. Pupeza, T. Eidam, B. Bernhardt, A. Ozawa, J. Rauschenberger, E. Fill, A. Apolonski, Th. Udem, J. Limpert, Z. A. Alahmed, Abdallah M Azzeer, T. W. Hänsch, A. Tünnermann, F. Krausz, "Power Scaling of a 78 MHz-Repetition Rate Femtosecond Enhancement Cavity," CLEO (2010).
- I. Pupeza, T. Eidam, J. Kaster, B. Bernhardt, J. Rauschenberger, A. Ozawa, E. Fill, Th. Udem, M. F. Kling, J. Limpert, Z. A. Alahmed, A. M. Azzeer, A. Tünnermann, T. W. Hänsch and F. Krausz, "Power Scaling and High-Power Applications of a Femtosecond Enhancement Cavity," invited talk, Photonics West (2011).
- I. Pupeza, J. Kaster, S. Holzberger, T. Eidam, B. Bernhardt, E. Fill, V. Pervak, A. Apolonski, Th. Udem, R. Holzwarth, J. Limpert, Z. A. Alahmed, A. M. Azzeer, A. Tünnermann, T. W. Hänsch and F. Krausz, "Generation of Brilliant XUV Radiation with Enhancement Cavities," invited talk, SIECPC (2011).
- I. Pupeza, J. Kaster, T. Eidam, B. Bernhardt, R. Holzwarth, T. W. Hänsch, Th. Udem, J. Limpert, A. Tünnermann, E. Fill, F. Krausz, "Progress in Enhancement Cavities for XUV Generation," CLEO (2011).
- I. Pupeza, J. Weitenberg, P. Rußbüldt, T. Eidam, J. Limpert, E. Fill, Th. Udem, H.- D. Homann, R. Poprawe, A. Tünnermann, F. Krausz, "Tailored Transverse Modes in High-Finesse Femtosecond Enhancement Cavities," CLEO (2011).

Patent Applications

- F. Krausz, E. Fill, J. Rauschenberger, I. Pupeza, "Method and laser device for generating pulsed high power laser light," Patent Application number PCT/EP2009/008278, (2009).
- J. Kaster, I. Pupeza, E. Fill, F. Krausz, "Method of generating enhanced intraresonator laser light, enhancement resonator and laser device," Patent Application number PCT/EP2010/005464, (2010).
- J. Weitenberg, P. Rußbüldt, Th. Udem, I. Pupeza, "Optischer Resonator mit direktem geometrischem Zugang auf der optischen Achse," Patent Application number 102011008225.5, (2011).
- I. Pupeza, F. Krausz, "Spatially splitting or combining radiation beams," Patent Application number 11000626.9, (2011).